Essentials of nonlinear control theory

J. R. Leigh

PETER PEREGRINUS LTD
on behalf of the
Institution of Electrical Engineers

Published by Peter Peregrinus Ltd., London, UK.

©1983: Peter Peregrinus Ltd

British Library Cataloguing in Publication Data

Leigh, J.R.
   The essentials of nonlinear control theory.
   —(IEE topics in control series; 2)
   1. Control theory   2. Nonlinear theories
   I. Title   II. Series
   629.8'36   QA402.3

ISBN 0-906048-96-6

Printed in England by Short Run Press Ltd., Exeter

Preface

The book gives a concise coverage of what might be termed classical nonlinear control theory. It contains the essentials for a course at either final year undergraduate or first year graduate level and most of the material has been class-tested with these categories of students at the Polytechnic of Central London.

The references and selective bibliography allow the reader to consolidate his understanding and to pursue aspects of the subject to its current boundaries.

Acknowledgement

I record my thanks to Mr. M. H. Ng who computed the examples in Chapter 11.

List of symbols

Elements within a vector x are denoted x_i.

s	the complex variable of Laplace transformation
$\dot{x}$	the time derivative of x
$\tilde{x}$	the critical point x
$A: X \to Y$	the mapping from space X to space Y
$I(x)$	the imaginary part of x
$\mathcal{R}(x)$	the real part of x
$\mathbf{R}^n$	real n-dimensional space
π, π^+, π^-	trajectories (see Section 2.2)
∇x	gradient of the vector x
$\ln$	natural logarithm
$(\ )^T$	transpose
$\forall$	for all, for every
$\|\ \|$	norm
$\triangleq$	equal to by definition
$\langle\ ,\ \rangle$	inner product
$\to$	implies
$\{x(t) \mid x(0) = 0\}$	the set of functions $x(t)$ such that $x(0) = 0$

Contents

Chapter 1

Initial orientation

1.1 Origins of nonlinearity

Every signal found in a real system must have an upper bound and if system operation causes the upper bound to be reached, then linearity is lost. This type of nonlinearity is commonly called saturation.

Electrical machines and transformers contain ferro-magnetic cores, causing highly nonlinear magnetisation curves to appear within the system equations.

Mechanical devices, such as gearboxes, typically exhibit backlash so that, on application of a very small signal, there may be no output. Most practical mechanical devices are subject to highly nonlinear friction forces.

Many control systems incorporate relays to achieve fast response at low cost. Such systems are, of course, highly nonlinear.

Very many of the basic equations governing, for instance, chemical, thermal and fluid processes are inherently nonlinear, so that control of such processes may require nonlinear techniques. Biological processes — an increasingly important class — are also usually inherently nonlinear.

1.2 Definition of nonlinearity

A function f is linear if:

(a) $f(u_1 + u_2) = f(u_1) + f(u_2)$ for any u_1, u_2 in the domain of f;

(b) $f(\alpha u) = \alpha f(u)$ for any u in the domain of f and for any real number α.

These equivalences are illustrated in Fig. 1.1. Every other function is nonlinear. Thus, there is no unity among nonlinear functions or nonlinear systems — they have in common only the absence of a particular property.

1.3 Types of behaviour found in nonlinear systems

Certain types of behaviour are possible only in nonlinear systems. Some examples are quoted below.

A nonlinear system, subject to a sinusoidal input at a frequency ω, may have harmonics at frequencies $k\omega$, $k = 2, 3, \ldots$ present in the output. Subharmonics at frequencies less than ω may also be present. If two sinusoids at frequencies ω_1, ω_2 are input simultaneously then the output will contain components at frequencies $\omega_1 + \omega_2$ and $\omega_1 - \omega_2$.

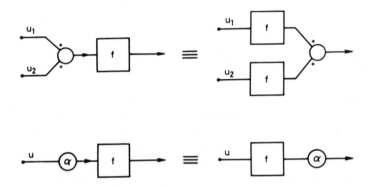

Fig. 1.1 *The equivalences that define linearity*

A nonlinear system may exhibit qualitatively different types of behaviour in different regions, according to choice of initial conditions. It is possible to imagine a number of 'cells'. Within each cell one type of behaviour occurs. Clearly, it would be the goal of analysis to determine, at least approximately, the boundaries of the cells for a particular system. (For a linear system, only one type of behaviour is possible. There is one cell extending to infinity.) Stability is no longer simple to define, since a solution may grow in one region and then come to equilibrium in another region.

A nonlinear system may oscillate stably at a particular amplitude and frequency, in a manner that is impossible in a linear system. This occurs when the system possesses a limit cycle — or unique periodic trajectory. One cause of such behaviour in a mechanical system is nonlinear friction, giving rise to so-called stick-slip motion.

Jump resonance is another, essentially nonlinear, phenomenon. It can occur in a closed loop containing a saturation type nonlinearity (Fig. 1.2) and can be understood by reference to Fig. 1.3. While the frequency is increasing, the system gain follows the curve ABDEF, whereas, while the frequency is decreasing, it follows the curve FECBA. In both directions, the system gain makes an instantaneous change or jump.

1.4 Failure of linear techniques

The absence of linearity makes system analysis difficult, since so many standard control techniques fail.

A linear system subjected to a sinusoidal input produces a sinusoidal output at the same frequency. Frequency response methods (for example the graphical methods of Nyquist, Bode and Nichols) rely on this property and hence are not applicable to nonlinear systems.

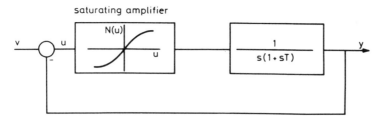

Fig. 1.2   *A closed-loop containing a saturating amplifier*

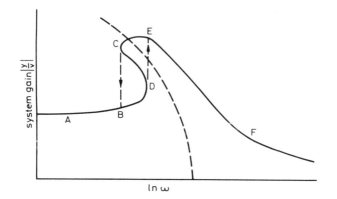

Fig. 1.3   *The jump resonance phenomenon*

Linear systems obey superposition principles allowing, in the limit, calculation of system outputs by the method of convolution. For nonlinear systems, superposition is invalid by definition. Transfer functions and techniques based on Laplace transforms are therefore not applicable. Pole-zero diagrams and root-locus techniques depend on transfer functions and are also inapplicable. Step-response methods are also of limited usefulness, because of the amplitude dependence of nonlinear systems.

Block diagrams and signal flow graphs can be drawn for nonlinear systems but they cannot be manipulated and simplified by block-diagram algebra or by Mason's rules.

Nonlinear differential equations follow no standard pattern, cannot in general be expressed in vector-matrix form, nor be usefully expressed in terms of the operator D. Consequently, such equations cannot be tested for stability by Hurwitz or Routh methods or by eigenvalue techniques. Rarely do practically useful nonlinear differential equations have a closed form solution.

1.5 Approaches to nonlinear system analysis

One group of approaches involves replacement of the nonlinear system by (in some sense) a linear approximation. Analysis of the linear system then gives information about the original nonlinear system. The describing function method and Lyapunov's first method are in this group.

A second approach attempts to find useful information about a system without solving the system equations. Lyapunov's second or direct method has such an approach to determine approximate regions of stability.

A third approach uses what I have called 'envelope techniques'. Here, the nonlinearity is enveloped graphically in a linear segment and a 'worst-case' stability analysis is then possible.

Numerical solution of nonlinear equations is, of course, always possible but is of limited usefulness in obtaining an overall insight into system behaviour.

A useful graphical technique for the display of nonlinear system behaviour is the phase-plane. Despite the fact that the display is virtually restricted to second-order systems, it allows most of the important concepts of nonlinear analysis to be illustrated.

Chapter 2

Dynamic systems — definitions and notation

2.0 Introduction

The book is concerned with the study of nonlinear dynamic systems. A dynamic system will often be denoted by the symbol Σ, a convenient abbreviation that can be given a precise mathematical meaning when necessary. The systems to be considered will have a finite number n of state variables, denoted $x_1, \dots, x_n$. The vector $x = (x_1, \dots, x_n)^T$ will be an element in the real n-dimensional state space X and such an n-dimensional space in general will be denoted by the symbol R^n. We say that x is an element of the space X or more briefly $x \in X$. Time t, in a continuous time system, can be considered to satisfy the inequality $-\infty < t < \infty$. Alternatively, we can say that $t \in T = R^1$.

2.1 Dynamic system axioms

In the case where the system Σ is time invariant and receives no input, it is clear that $x(t)$ must be determined by $t, x(0)$ and Σ. We can think of Σ as a mapping and write

$$\Sigma: T \times X \to X$$

or

$$\Sigma(t, x_0) = x(t). \tag{2.1}$$

$x(t)$ is called a *particular solution* for the system Σ.

In order to be a well posed *dynamic system* (i.e., possessing a unique solution that depends continuously on its arguments), Σ must satisfy the axioms:

(*a*) Σ must be a continuous map.
(*b*) Σ must satisfy the condition

$$\Sigma(t, \Sigma(s, x)) = \Sigma(t + s, x), \quad t, s \in T.$$

This is called the *semi-group axiom* — it is clearly necessary if solutions are to be unique.

(c) $\Sigma(0, x) = x$, the *identity property*.

In the subject of topological dynamics, Σ is referred to as a *continuous flow*.

2.2 Trajectories and critical points

The *trajectory* through any point $y \in X$ is denoted $\pi(y)$ and is defined by

$$\pi(y) = \{x(t)|x(0) = y, \quad -\infty < t < \infty\}.$$

The *positive semi-trajectory* π^+ through y is defined

$$\pi^+(y) = \{x(t)|x(0) = y, \quad 0 < t < \infty\}$$

and the *negative semi-trajectory* π^- by

$$\pi^-(y) = \{x(t)|x(0) = y, \quad -\infty < t < 0\}.$$

A point $\tilde{x} \in X$ is called a *critical point* for the system Σ if

$$\Sigma(t, \tilde{x}) = \tilde{x} \quad \text{for every } t \in R^1.$$

2.3 Definitions of stability

A critical point $\tilde{x}$ of the system Σ is:

(a) *stable* if, given a circular region of radius $\delta > 0$ around the critical point, there exists another circular region, of radius ϵ, concentric with the first region, such that every positive half trajectory starting in the δ region remains within the ϵ region.

(b) *asymptotically stable* if it is stable and if in addition, every half trajectory satisfying the conditions in (a), reaches the critical point in the limit as $t \to \infty$.

A solution $x(t)$, originating at x_0, t_0 is stable with respect to the critical point $\tilde{x}$ if:

(a) $x(t)$ is defined for all t satisfying $t_0 < t < \infty$.

(b) If $\| x_0 - \tilde{x} \| < \delta$, for some positive constant δ, then there exists another positive constant ϵ such that $\|x(t) - \tilde{x}\| < \epsilon$, $\forall t \in (t_0, \infty)$

Figure 2.1 illustrates the definition of stability.

A solution is *asymptotically stable* to the critical point $\tilde{x}$ if it is stable and in addition

$$\lim_{t \to \infty} \|x(t) - \tilde{x}\| = 0.$$

A solution originating at x_0, t_0 is *uniformly stable* if it is stable for any choice t_0' of initial time, $t_0' \geq t_0$, the constants δ, ϵ being fixed. (The concept of uniform stability is required when working with time varying systems.)

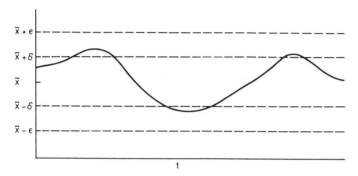

Fig. 2.1 *Illustration of the definition of stability*

2.4 Representation of a system by differential equations

The most common concrete form for representation of a time invariant system Σ is the finite set of ordinary differential equations

$$\dot{x}_i = f_i(x_1, \ldots , x_n), \quad x_i(0) = x_{i0}, \quad i = 1, \ldots , n,$$

which can be written more concisely in the form

$$\dot{x} = f(x), \qquad x(0) = x_0, \tag{2.2}$$

where $x \in X = \mathsf{R}^n$ and f is a vector valued mapping $f: X \to X$.

Provided that restrictions corresponding to the axioms (a) to (c) of Section 2.1 are imposed on the differential equations, the system Σ and its differential equation representation can be regarded as equivalent.

In control applications, it will rarely be required to solve a nonlinear differential equation analytically. More usually, information will be sought on the types of solution, on the boundaries between these types of solution, on stability properties, and so on.

It is useful to have some mathematical background on differential equations, since they form the primary models of nonlinear systems. A brief discussion follows. It is sufficient for the purposes of this book but the reader might with advantage consult a text on differential equations. Driver (1977) and Braun (1975) are particularly recommended.

2.5 First-order scalar differential equations

The general first-order equation has the form

$$\dot{x} = f(t, x), \quad x(t_0) = x_0. \tag{2.3}$$

f is a given real-valued function defined on the region D where

$$t \in (\alpha, \beta) \quad \text{and} \quad x \in (\gamma, \delta).$$

A solution of eqn. (2.3) is a continuous function on an interval $J = (\alpha_1, \beta_1)$ such that

(a) $t_0 \in J$;
(b) $x(t_0) = x_0$;
(c) For all t in J, $(t, x(t)) \in D$ and $\dot{x} = f(t, x)$.

Notes:
(a) $x(t)$ must be differentiable on J.
(b) If f is continuous then $\dot{x}(t)$ must also be continuous.

2.5.1 *Theorem 2.1*
Let f and x be continuous functions and let $(t, x(t)) \in D$ for all t in J then x is a solution of eqn. (2.3) if and only if

$$x(t) = x_0 + \int_{t_0}^{t} f[p, x(p)] \, dp \quad \text{for all } t \text{ in } J.$$

Proof: Differentiate the last expression to obtain eqn. (2.3).

2.5.2 *Definition*
The function f is said to satisfy a *Lipschitz condition* on J if for some constant k

$$|f(t,x_1) - f(t,x_2)| \leq k|x_1 - x_2|$$

whenever $t \in J$ and for any x_1, x_2.

2.5.3 *Theorem 2.2*
Let f be continuous and satisfy a Lipschitz condition everywhere on J. Then eqn. (2.3) has a solution on the inverval (α, β). The solution is unique and computable (i.e., it can be determined by a successive approximation method)

Proof: See Driver (1977).

2.6 Systems of first-order equations

By introducing dummy variables, a differential equation of order n can be expressed as n first-order equations in the form of eqn. (2.2). For example, the equation

$$\ddot{\theta} + a\dot{\theta} + b\theta = u$$

can be so expressed by the substitutions

$$x_1 = \theta, \qquad x_2 = \dot{x}_1,$$

yielding

$$\left.\begin{array}{l} \dot{x}_1 = x_2, \\ \dot{x}_2 = -bx_1 - ax_2 + u. \end{array}\right\}$$

Since any combination of differential equations can be transformed into the form of eqn. (2.2), it can be seen that the representation is of great generality. The conditions that guarantee the existence of unique solutions of finite sets of first-order differential equations are reasonably straightforward analogues of the conditions given earlier for a single scalar equation.

If in eqn. (2.2) there exists a point $\bar{x}$ for which $f(\bar{x}) = 0$, then clearly $\bar{x}$ must be a critical point of the corresponding system Σ.

2.7 Vector field aspects

The function f can sometimes be regarded usefully as being the generator of a vector field that is everywhere tangential, except at critical points, to the trajectories of the system.

A system Σ is said to be *conservative* if the total energy $V(t)$ is constant for all values of t. For Σ to be a conservative system, the vector function f in eqn. (2.2) is required to satisfy the condition

$$\frac{\partial f_1}{\partial x_1} + \frac{\partial f_2}{\partial x_2} + \ldots + \frac{\partial f_n}{\partial x_n} = 0.$$

2.8 Limit cycles

Any trajectory that is closed on itself must clearly represent a periodic solution. Let C be such a closed trajectory that is *isolated*, i.e., is not one of a continuous family of similar closed trajectories. Then C is called a *stable limit cycle*. Conversely, if in the neighbourhood of C, solutions tend to C with decreasing time, C is called an *unstable limit cycle*.

Limit cycles are very important in the theory of oscillators. They are discussed more fully in Chapter 9.

Chapter 3

The describing function method

3.0 Introduction

The describing function method can be considered as an extension to nonlinear systems of the usual Nyquist stability criterion. The origins of the method can be traced back to nonlinear mechanics (Krylov and Bogoliuboff, 1932) but the first workers to apply the approach specifically to control systems were Tustin (1947) and Kochenburger (1950).

3.1 Outline of the method

Suppose that a nonlinear feedback loop can be represented as in Fig. 3.1, in which $G(s)$ is a linear transfer function and f is a nonlinear function, defined either analytically or graphically.

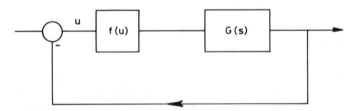

Fig. 3.1 *A closed-loop containing a nonlinear function f*

Assume that a sinusoidal signal $u = a \sin \omega t$ is input to the nonlinear function f. By Fourier methods the corresponding output from the function can be expressed

$$f(a \sin \omega t) = a_0 + \sum_{k=1}^{\infty} (a_k \cos k \omega t + b_k \sin k \omega t).$$

The output of the nonlinear element consists of a fundamental component $a_1 \sin \omega t + b_1 \cos \omega t$ together with a mean level a_0 and harmonic components at frequencies 2ω, 3ω, ...

The describing function method relies on neglecting all but the fundamental component. The nonlinear function f can then be approximated by its 'equivalent gain' denoted $N(a)$ and defined by the expression

$$N(a) = \frac{a_1 \sin \omega t + b_1 \cos \omega t}{a \sin \omega t} = \frac{a_1}{a} + j \frac{b_1}{a}.$$

$N(a)$, as the notation indicates, is a complex number depending on the amplitude a of the input sinusoid. $N(a)$ is conveniently displayed as a locus in the complex plane.

Considering only the fundamental component ω to be significant, the steady-state sinusoidal behaviour of the closed loop is described by the equation

$$G(j\omega)N(a) = -1,$$

or

$$G(j\omega) = \frac{-1}{N(a)}.$$

The stability of the loop can be analysed directly by Nyquist's criterion, with the locus of $-1/N(a)$ being considered as a generalisation of the $(-1, 0)$ point of the linear case. Thus, encirclement of the $-1/N(a)$ locus by the $G(j\omega)$ locus indicates instability of the closed loop. Intersection between the two loci indicates the possibility of continuous oscillation. No encirclement by the $G(j\omega)$ locus of the $-1/N(a)$ locus and no intersection between the loci indicates a stable non-oscillatory solution.

The validity of the method rests on the validity of the assumption that, although the nonlinearity f produces harmonics, they can be neglected. Numerical examples to be given later will confirm that the assumption is valid, provided that $G(s)$ behaves like a low-pass filter.

Before proceeding further, we summarise the method.

(*a*) It is applicable to the analysis of single-loops without input, in which the nonlinearity can be represented by a function f (defined analytically or graphically) in series with a linear transfer function $G(s)$.

(*b*) A sinusoidal signal $a \sin \omega t$ is assumed to be the input to f. The equivalent gain $N(a)$ of f is then defined to be

$$N(a) = \frac{\text{fundamental component in output}}{\text{input}}.$$

(*c*) The system is governed by the equation $G(j\omega) = -1/N(a)$. Every point on the $-1/N(a)$ locus in the complex plane plays the part of the $(-1, 0)$ point in the linear case. By plotting the $G(j\omega)$ and $-1/N(a)$ loci in the complex plane, the stability of the loop can be assessed.

(*d*) The assumption that harmonics can be neglected will be true provided that $G(s)$ behaves as a low-pass filter.

3.2 Justification for the describing function method

It is usual to justify the method by plausible physical reasoning (as above), rather than by mathematical proof, but see Bass (1961) for a rigorous proof of the method. It is sufficient for most practical cases to have an approximate knowledge of the magnitudes of the terms that are being neglected. Such knowledge can easily be obtained during the Fourier analysis of the output of the nonlinear function.

3.3 Stability of self-excited oscillations

As stated earlier, the method predicts the possibility of continuous oscillation whenever the $G(j\omega)$ and $-1/N(a)$ loci intersect. However, such oscillations may be stable or unstable — only stable oscillations will exist in practice. To determine the stability of self-excited oscillations the following approach will nearly always suffice. Consider an intersection p of the $G(j\omega)$ and $-1/N(a)$ loci and assume a small decrease in amplitude a. The representative point on the $-1/N(a)$ locus will move to a new point. If the new point is encircled by the $G(j\omega)$ locus, the oscillations will grow, the system will tend to return to the original intersection p and the oscillations are stable. Conversely, if the new point is not encircled, the system will move further and further from the intersection and the oscillations are unstable.

From this argument, the following simple geometric criterion follows. Let p be an intersection of the $G(j\omega)$ and $-1/N(a)$ loci. Let x, y be the gradients of the two loci respectively at p. Then, if the angle θ that y makes with x is less than 180°, the predicted self-excited oscillations will be stable. If the angle θ is more than 180°, the oscillations will be unstable. Figure 3.2 illustrates the concept. Intersections p_1 and p_2 represent stable and unstable oscillations respectively.

The amplitude a and frequency ω of self-excited oscillations can be estimated from the $N(a)$ and $G(j\omega)$ loci, respectively, at the points where they intersect.

3.4 Derivation of the Fourier series for $f(a \sin \omega t)$

The amount of calculation can be reduced by noting how the Fourier coefficients depend on the function f:

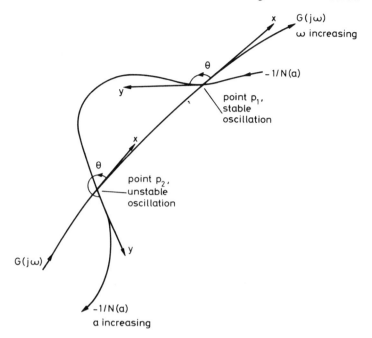

Fig. 3.2 *Stable and unstable oscillations predicted by the describing function method*

(a) If f is single-valued (i.e., to each a corresponds only one value $f(a)$), then $b_2, b_4, b_6, \ldots = 0$ and $a_1, a_3, a_5, \ldots = 0$.

(b) If f is single-valued and skew-symmetric (i.e., such that $f(u) = -f(-u)$), then $b_2, b_4, b_6, \ldots = 0$ and all a coefficients are zero.

(c) If f is skew-symmetric but not single valued then all even coefficients, including the coefficient a_0, are zero.

If f is single-valued, skew-symmetric and a polynomial of up to power three then

$$N(a) = \frac{2}{3a}(f(a) + f(a/2)). \tag{3.1}$$

The formula was derived by Tsypkin (1958), in a paper containing other formulae for higher order polynomial functions.

Equation (3.1) can be used with advantage to obtain an approximate solution for any single-valued skew-symmetric function.

3.5 Determination of Fourier coefficients without integration

Let $f(t)$ be any function, of period 2π, that becomes identically zero after a finite number of differentiation operations.

By a *jump j* in the function at a point t_0 we mean the difference between left and right limits at t_0, i.e.,

$$j_{x_0} = f(x_0 + \epsilon) - f(x_0 - \epsilon) \quad \text{as } \epsilon \to 0.$$

Jumps in the first, second, third and higher derivatives at x_0 are denoted by

$$j'_{x_0}, j''_{x_0}, j'''_{x_0}, \cdots$$

It is shown in Kreyszig (1974) that the Fourier coefficients $a_i, b_i, i = 1, \ldots$ (but not a_0) representing f can be determined from the formulae

$$a_n = \frac{1}{n\pi} \left[-\sum_{s=1}^{m} j_s \sin(nx_s) - \frac{1}{n}\sum_{s=1}^{m} j'_s \cos(nx_s) + \frac{1}{n^2}\sum_{s=1}^{m} j''_s \sin(nx_s) \right.$$

$$\left. + \frac{1}{n^3}\sum_{s=1}^{m} j'''_s \cos(nx_s), - - + + \right],$$

$$b_n = \frac{1}{n\pi} \left[\sum_{s=1}^{m} j_s \cos(nx_s) - \frac{1}{n}\sum_{s=1}^{m} j'_s \sin(nx_s) - \frac{1}{n^2}\sum_{s=1}^{m} j''_s \cos(nx_s) \right.$$

$$\left. + \frac{1}{n^3}\sum_{s=1}^{m} j'''_s \sin(nx_s), + - - + \right],$$

where each summation in s runs through the m values at which jumps occur.

3.5.1 Simple example

$$f(t) = -1, \quad -\pi < t < 0,$$
$$f(t) = 1, \quad 0 < t < \pi.$$

The jumps are $j_1 = 2$ at $t = 0$, $j_2 = -2$ at $t = \pi$. (The jump at $-\pi$ is considered to belong to the previous cycle.)

There are no further jumps to be taken into account, therefore,

$$b_n = \frac{1}{n\pi} [2 \cos 0 - 2 \cos n\pi] = \frac{2}{n\pi} [1 - \cos n\pi].$$

The a coefficients are zero because f is an odd function.

The method can save time in the calculation of describing functions for relay type nonlinearities. In those applications where the nonlinearity passes part of the applied sinusoid through to the output, the method is clearly inapplicable.

3.6 Examples

3.6.1 Ideal relay (Fig. 3.3)
The function $f(u)$ obeys the rule

$$f(u) = 1, \qquad u > 0,$$

$$f(u) = -1, \qquad u < 0.$$

In response to an input $u = a \sin \omega t$, the function produces a rectangular wave of unit amplitude, in phase with the input signal. From the skew-symmetry and single valuedness of f it is known that the only non-zero Fourier coefficients are $b_1, b_3, b_5, \ldots$ and that the Fourier coefficient b_1 represents the magnitude of the fundamental component in the output from f.

$$b_1 = \frac{2}{\pi} \int_0^\pi 1 \sin \omega t \, d\omega t = \frac{4}{\pi} \ .$$

Hence, the fundamental component in the output of f is $4 \sin \omega t / \pi$. $N(a) = (4 \sin \omega t / \pi)/(a \sin \omega t) = 4/\pi a$.

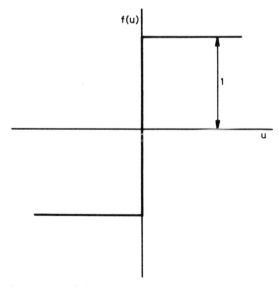

f(u)

1

u

Fig. 3.3   *Ideal relay characteristic*

Letting a take on all positive values results in the locus of $-1/N(a) = -a\pi/4$ occupying the whole of the negative real axis in the complex plane (Fig. 3.4). It is immediately apparent that any closed loop containing an ideal relay will experience continuous self-excited oscillations in every case where the open-loop locus $(G(j\omega))$ of the linear dynamic part crosses the negative real axis in the complex plane. The frequency ω^* of such oscillations can be found by solving the equation

$$\phi(G(j\omega^*)) = -180°,$$

where ϕ represents the phase angle of the transfer function G. The amplitude a^* of the oscillations then follows from the equation

$$a^* = \frac{4}{\pi}|G(j\omega^*)|.$$

The amplitude of the third harmonic in the relay output is given by

$$b_3 = \frac{2}{\pi}\int_0^\pi 1 \sin 3 \,\omega t \,d\omega t = \frac{4}{3\pi} \,.$$

The amplitude of the third harmonic is one third of the amplitude of the fundamental. Suppose $G(s)$ is an integrator then, after passing through $G(s)$, the amplitude of the third harmonic will be one ninth of the amplitude of the fundamental. Higher odd harmonics will be of smaller amplitudes.

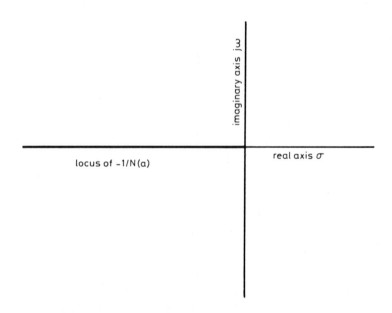

Fig. 3.4   *The – 1/N(a) locus for the ideal relay*

3.6.2 Example 2. Relay with dead-zone (Fig. 3.5)
The output of the function f in response to an input $u = a \sin \omega t$ is a rectangular wave (Fig. 3.6). Again, $a_1 = 0$ from the form of f.

$$b_1 = \frac{2}{\pi} \int_\alpha^{\pi-\alpha} k \sin \omega t \, d\omega t \stackrel{+}{-} \frac{4k}{\pi a} (a^2 - d^2)^{1/2},$$

where

$$\alpha = \sin^{-1} d/a,$$

$$-\frac{1}{N(a)} = \frac{-a\pi}{4k} \frac{1}{(1 - (d/a)^2)^{1/2}}.$$

Letting a range over all positive values, results in the locus $- 1/N(a)$ moving along the negative real axis in the complex plane from $- \infty$ to a point $- p$ and then returning to $- \infty$. Since $- p$ is a turning point on the locus, it can be found by differentiation. Calculation yields the result $p = d\pi/2k$. As expected, the result shows that self-oscillation can be prevented provided that a large enough dead-zone d is introduced to allow the $G(j\omega)$ locus to avoid the $- 1/N(a)$ locus (Fig. 3.7).

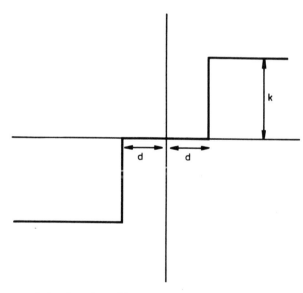

Fig. 3.5 *Characteristic of a relay with dead-space*

3.6.3 Example 3. Cubic nonlinearity (Fig. 3.8)

$$b_1 = 3a^3/4, \quad -\frac{1}{N(a)} = -4/3a^2.$$

Again, the $- 1/N(a)$ locus occupies the whole negative real axis. The third harmonic amplitude $b_3 = - a^3/4$.

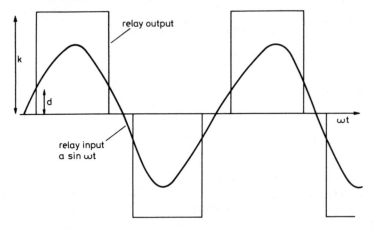

Fig. 3.6   *Response of the relay with dead-space to a sinusoidal input*

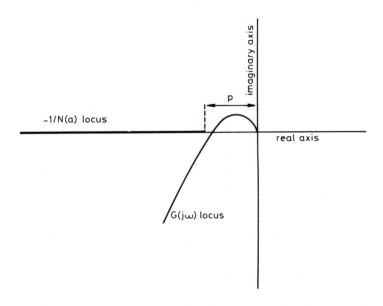

Fig. 3.7   *The – 1/N(a) locus for the relay with dead-space showing how the introduction of dead-space may prevent oscillation*

3.6.4 Example 4.

The nonlinear element has the form shown in Fig. 3.9. In response to an input $a \sin \omega t$ it has the output

$$ca \sin \omega t + d, \quad 0 \le \omega t \le \pi/2,$$

$$ca \sin \omega t - d, \quad \pi/2 < \omega t \le \pi.$$

Fitting Fourier coefficients,

$$a_1 = \frac{2}{\pi}\left[\int_0^{\pi/2} (ca \sin \omega t + d) \sin \omega t \, d\omega t\right.$$

$$\left. + \int_{\pi/2}^{\pi} (ca \sin \omega t = d) \cos \omega t \, d\omega t\right] = ca.$$

$$b_1 = \frac{2}{\pi}\left[\int_0^{\pi/2} (ca \sin \omega t + d) \cos \omega t \, d\omega t + \int_{\pi/2}^{\pi}\right.$$

$$\left. (ca \sin \omega t - d) \cos \omega t \, d\omega t\right]$$

$$= \frac{2}{\pi}\left[d \sin \omega t \Big|_0^{\pi/2} - d \sin \omega t \Big|_{\pi/2}^{\pi}\right]$$

$$= 4d/\pi.$$

$N(a)$ is given by

$$N(a) = \frac{a_1}{a} + j\frac{b_1}{a} = c + j \, 4d/\pi a.$$

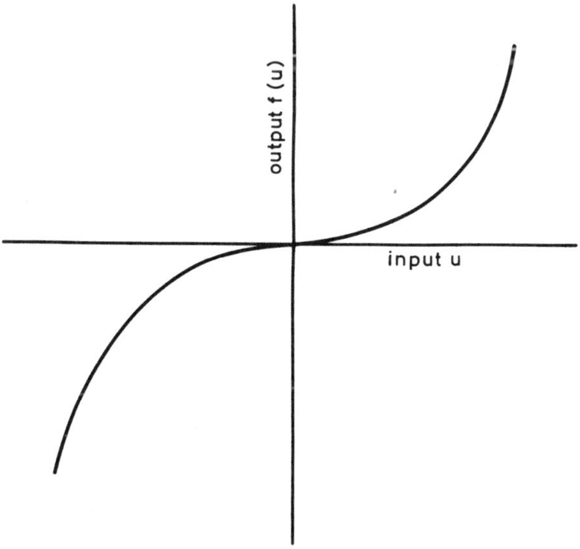

Fig. 3.8 *Cubic nonlinearity*

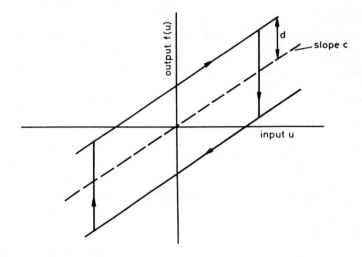

Fig. 3.9   *Nonlinear characteristic for Example 4*

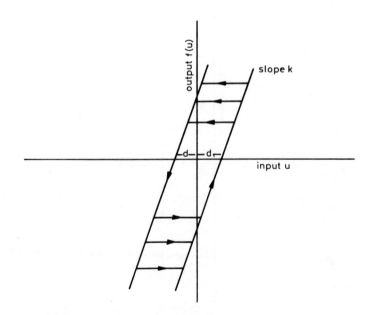

Fig. 3.10   *Nonlinear characteristic for Example 5*

3.6.5 *Example 5. Linear element with hysteresis (Fig. 3.10)*

The function f has the output

$$y = k(u - d), \qquad \dot{u} > 0,$$
$$y = k(u + d), \qquad \dot{u} < 0.$$

When $\dot{u}$ changes sign, y remains constant until one or other of the two characteristics is reached. The hysteresis introduces phase shift and the describing function is complex valued. Calculation gives

$$N(a) = \frac{k}{2} + \frac{k}{\pi} (\sin^{-1}\alpha + 2\alpha\beta^{1/2}) - j \left(\frac{4k\beta}{\pi}\right),$$

where

$$\alpha = 1 - 2d/a, \quad \beta = (d/a)(1 - d/a).$$

The $-1/N(a)$ locus has the form shown in Fig. 3.11.

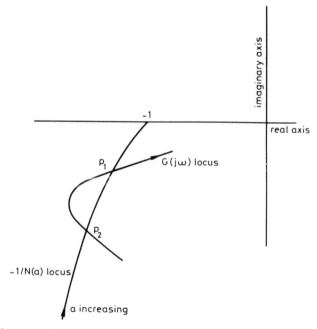

Fig. 3.11 *Interpretation of Example 5 in the complex plane*

Suppose the $G(j\omega)$ locus of the linear element in series with f is as shown in the figure, then at the two intersections p_1 and p_2, continuous oscillations are possible. Applying the rule for determining the stability of continuous oscillations, it is found that $\theta(p_1) > 180°$ and $\theta(p_2) < 180°$. Thus, stable self-excited oscillations will occur at the point p_2, but not at p_1.

3.7 Concluding remarks

An extensive literature exists on the describing function method and many extensions to the basic method have been made. In particular:

(*a*) Systems with input can also be analysed Grensted (1962). In this case, possible complications due to ultra-harmonic resonance or sub-harmonic resonance have to be considered.

(*b*) Booton (1954) developed a 'statistical describing function' for the stochastic analysis of loops containing nonlinearity. The approach is analogous to that of the usual describing function except that the signal to the nonlinearity is assumed to be Gaussian. The resulting output is non-Gaussian, due to the nonlinearity, but the assumption is made that the non-Gaussian elements can be neglected (analogous to the neglections of harmonics in the usual describing function method).

(*c*) It is possible to derive the describing function $N(a)$ for a particular nonlinearity experimentally with the aid of a transfer function analyser. Should the need arise, the function f can be derived from $N(a)$ by a method due to Zadeh (1956).

Chapter 4

The phase plane portrait

4.0 Introduction

Since frequency response techniques and root locus diagrams are not applicable to a nonlinear process, there is an important need for a graphical tool to allow nonlinear behaviour to be displayed. This need is filled by the phase plane diagram. The method is applicable to second-order processes without input, although effects equivalent to step or ramp inputs can be obtained by choice of initial conditions.

4.1 Definition

Given a second-order process, written in the form

$$\dot{x}_1 = f_1(x_1, x_2),$$
$$\dot{x}_2 = f_2(x_1, x_2).$$

The *phase plane* consists of plots of trajectories in the x_1, x_2 plane. The phase plane with specimen trajectories indicating the nature of the solutions is called a *phase-portrait*.

4.2 Simple examples illustrating behaviour in the phase plane

4.2.1 A second-order linear process without damping
Given the equation $\ddot{y} + y = 0$, set $x_1 = y$, $x_2 = \dot{y}$ to yield $\dot{x}_2 + x_1 = 0$. Divide by $\dot{x}_1 = x_2$ to obtain

$$\frac{\dot{x}_2}{\dot{x}_1} + \frac{x_1}{x_2} = 0, \quad \frac{dx_2}{dx_1} + \frac{x_1}{x_2} = 0.$$

Integration gives

$$\frac{x_1^2}{2} + \frac{x_2^2}{2} = k,$$

where k is a constant of integration. The phase-portrait consists of concentric circles about the origin. The given initial condition determines the value of k and hence the radius of the trajectories.

4.2.2 A second-order linear process with damping

$\ddot{x} + k\dot{x} + x = 0$, $x(0)$, $\dot{x}(0)$ given. Put $x_1 = x$, $x_2 = \dot{x}_1$. The equation can then be written

$$\left. \begin{array}{l} \dot{x}_1 = x_2 \\ \dot{x}_2 = kx_2 - x_1 \end{array} \right\} x_1(0), x_2(0) \text{ given.} \tag{4.1}$$

Then

$$\frac{dx_2}{dx_1} = -k - x_1/x_2$$

$$x_2 = \frac{-x_1}{(dx_2/dx_1) + k}.$$

Suppose now that $k = 1$, then:

(a) on the x_2 axis, $dx_2/dx_1 = -1$;
(b) on the x_1 axis, $dx_2/dx_1 = \infty$;
(c) on the line $x_2 = x_1$, $dx_2/dx_1 = -2$;
(d) on the line $x_2 = -x_1$, $dx_2/dx_1 = 0$.

Knowledge of the gradients of the solution curves allows a sketch to be made of the approximate behaviour in the phase plane (Fig. 4.1). Each solution is a logarithmic spiral, satisfying at any time the condition

$$x_1^2(t) + x_2^2(t) = ce^{\mu t},$$

where c depends on the initial conditions. The coefficient μ depends on the eigenvalues λ_1, λ_2 of the system according to the equation

$$\mu = 2R(\lambda_1) = 2R(\lambda_2).$$

The trajectories do not tend to a limiting direction. To see this let $\lambda_1 = a + jb$, $\lambda_2 = a - jb$. The separate equations for x_1 and x_2 have the form

$$x_1(t) = \sqrt{c}\, e^{at} \cos bt,$$

$$x_2(t) = \sqrt{c}\, e^{at} \sin bt.$$

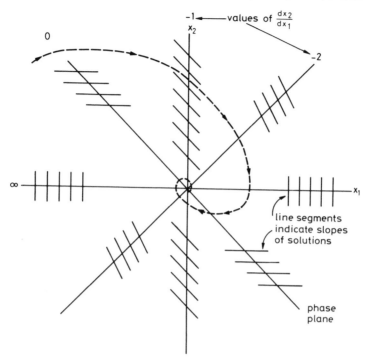

Fig. 4.1 *Phase plane portrait for Eqn. (4.1)*

The slope of the trajectories in the phase plane is given by $dx_2/dx_1 = \tan bt$ which does not tend to a limit as $t \to \infty$.

4.3 The direction of rotation of oscillatory trajectories in the phase plane

Consider a linear two-dimensional system $x = Ax$.

Any point on a trajectory can be expressed in polar coordinates in the form

$$|x|\underline{/\theta} = (x_1^2 + x_2^2)^{1/2}\underline{/\theta},$$

where

$$\theta = \tan^{-1}(x_2/x_1).$$

The direction of rotation is governed by

$$\dot{\theta} = \frac{d}{dt}(\tan^{-1}(x_2/x_1)) = \frac{1}{1 + (x_2/x_1)^2}\frac{x_1\dot{x}_2 - x_2\dot{x}_1}{x_1^2}.$$

The direction of rotation is clockwise if $\dot{\theta} < 0$, i.e., if $x, \dot{x}_2 - x_2\dot{x}_1 < 0$.

4.3.1 Example

$$\dot{x} = Ax \quad \text{with} \quad A = \begin{pmatrix} 0 & 1 \\ -1 & 0 \end{pmatrix}$$

Set

$$\dot{x}_1 = x_2, \quad \dot{x}_2 = -x_1,$$

$$x_1\dot{x}_2 - x_2\dot{x}_1 = -x_1^2 - x_2^2 < 0.$$

The solution rotates clockwise about the origin.

By considering the signs of $\dot{x}_1$ and $\dot{x}_2$ in particular regions, it is often possible to determine directions of rotation by inspection.

4.4 Switching lines in the phase plane

Consider the closed loop shown in Fig. 4.2. The relay produces an output $y(u)$ = 1 for $u > 0$ and $y(u) = -1$ for $u < 0$. The motor velocity x_2 is subject to the constraint $x_2 \leqslant |M|$, where M is a given constant. x_1 is the angular position of the motor shaft. A typical response is as shown in the phase plane diagram Fig. 4.3.

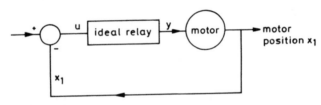

Fig. 4.2 *A position control-loop*

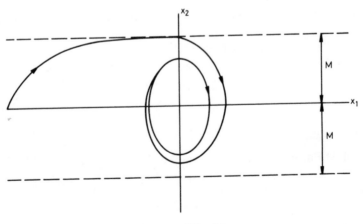

Fig. 4.3 *The response of the control-loop of Fig. 4.2*

The system can be stabilised by the addition of tachometer feedback (Fig. 4.4).

The input to the relay is

$$u = -x_1 - kx_2,$$

where k is the tachometer constant.

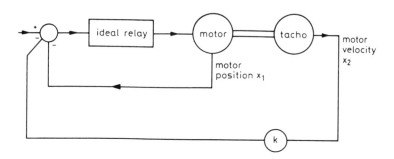

Fig. 4.4   *The loop of Fig. 4.2 with stabilising feedback*

The relay switches over when $-x_1 - kx_2 = 0$ and the line $x_2 = -x_1/k$ in the phase plane is called the switching line for the system. By variation of the tachometer constant k, the slope of the switching line and hence the degree of damping can be adjusted (Fig. 4.5).

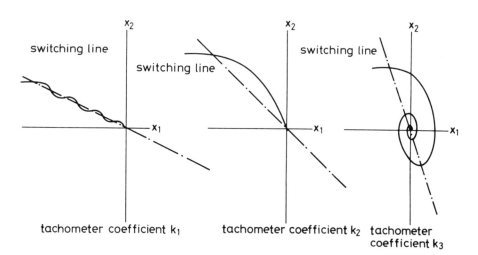

Fig. 4.5   *The response of the loop in Fig. 4.4 for different values of the tachometer feedback coefficient (where $k_1 > k_2 > k_3$)*

4.5 Discussion

The phase plane diagram will be used in a sketch-pad manner on many occasions. Further properties of systems in the phase plane are brought out in later sections of the book.

Note that the limitation to second-order systems is imposed largely to avoid the cumbersome problems of interpreting results graphically in more than two dimensions. Some results valid in the plane do not generalise immediately to higher dimensions. For instance, in the plane, every closed curve separates the plane into two regions. Clearly the result does not hold in higher dimensions.

Chapter 5

Linearisation

5.0 Introduction

The replacement of a nonlinear function by a linear approximation is known as linearisation. The motivation for linearisation is to allow analysis of a nonlinear problem by linear techniques. The results of such analysis need to be interpreted with care to ensure that the linearising approximation does not cause unacceptably large errors.

5.1 The principles of linearisation

Let f be an analytic function of one variable, then by Taylor's theorem,

$$f(a + h) = f(a) + h \left.\frac{df}{dh}\right|_a + h^2 \left.\frac{d^2f}{dh^2}\right|_a + \dots \,.$$

$\underset{\text{term}}{\text{constant}}$ $\underset{\text{term}}{\text{linear}}$ $\underbrace{\qquad\qquad}_{\text{higher order terms}}$

a is to be considered as the fixed point about which linearisation is performed whereas h is to be considered as a perturbation from the point a.

Linearisation consists in replacing the function $f(a + h)$ by the approximation

$$f(a + h) = f(a) + h \left.\frac{df}{dh}\right|_a.$$

The accuracy of the approximation depends on the magnitude of h and the magnitudes of the higher derivatives of f at $x = a$. Clearly, near to a turning-point of the function, $df/dh \simeq 0$ and the approximation will be ill-conditioned.

Where the function f operates on a vector, $f: R^n \to R^1$, the same principle, based on Taylor's theorem, still applies.

Let

$$a \in \mathbf{R}^n, \qquad h = \begin{pmatrix} h_1 \\ \vdots \\ h_n \end{pmatrix} \in \mathbf{R}^n,$$

then the appropriate linear approximation is given by

$$f(a + h) \simeq f(a) + h_1 \left. \frac{\partial f}{\partial h_1} \right|_a + \ldots + h_n \left. \frac{\partial f}{\partial h_n} \right|_a.$$

Geometrically, the action of linearisation is equivalent to putting a tangent plane onto the nonlinear surface f at the point a (see Fig. 5.1).

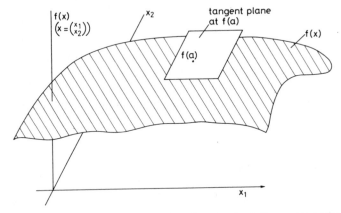

Fig. 5.1 *Illustration of a tangent plane forming the best linear approximation to a function f*

Notice particularly that in linearisation for control purposes, the point a does not need to be a single fixed point. It is often most useful to linearise about a trajectory. Such a trajectory may be specified analytically or it may be generated by successive approximations. Simple examples of linearisation follow.

5.1.1 Nonlinear second-order system with one input and two outputs

$$\dot{x}_1 = f_1(x_1, x_2, u),$$
$$\dot{x}_2 = f_2(x_1, x_2, u),$$
$$y_1 = g_1(x_1, x_2),$$
$$y_2 = g_2(x_1, x_2).$$

At a fixed point (x_1^*, x_2^*, u^*) the linear approximation is given in terms of Jacobian matrices as follows.

$$\begin{pmatrix} \dot{p}_1 \\ \dot{p}_1 \end{pmatrix} = \begin{pmatrix} \dfrac{\partial f_1}{\partial x_2} & \dfrac{\partial f_1}{\partial x_2} \\ \dfrac{\partial f_2}{\partial x_1} & \dfrac{\partial f_2}{\partial x_2} \end{pmatrix} \begin{pmatrix} p_1 \\ p_2 \end{pmatrix} + \begin{pmatrix} \dfrac{\partial f_1}{\partial u} \\ \dfrac{\partial f_2}{\partial u} \end{pmatrix}$$

$$\begin{pmatrix} r_1 \\ r_2 \end{pmatrix} = \begin{pmatrix} \dfrac{\partial g_1}{\partial x_1} & \dfrac{\partial g_1}{\partial x_2} \\ \dfrac{\partial g_2}{\partial x_1} & \dfrac{\partial g_2}{\partial x_2} \end{pmatrix} \begin{pmatrix} p_1 \\ p_2 \end{pmatrix}$$

where all derivatives are to be evaluated at

$$(x_1^*, x_2^*, u^*)$$

and where

$$p_1 = x_1 - x_1^*, \qquad p_2 = x_2 - x_2^*,$$
$$q = u - u^*,$$
$$r_1 = y_1 - g_1(x_1^*, x_2^*),$$
$$r_2 = y_2 - g_2(x_1^*, x_2^*),$$

i.e., the p, q, r variables represent perturbations from the steady operating condition.

A system that is governed by a control system to operate at a fixed point a is clearly a good candidate for linearisation at that point. The linear model then represents perturbations from the desired value.

5.2 Linearisation about an analytically specified trajectory

Here the process is given by $y = f(x)$ but x is supposed to vary along a nominal trajectory $x_s(t)$. Linearisation is to consist of considering perturbations from the nominal trajectory.

Suppose for concreteness that $f(x) = x^2$ and that $x_s = \sin t$, then the linear approximation will be given by

$$r(t) = (2 \sin t)\, p(t),$$

where

$$p(t) = x(t) - x_s(t),$$

and

$$r(t) = y(t) - (x_s(t))^2.$$

If now at $t = \pi/6$, $x = 0.52$ instead of the nominal value $x_s = 0.5$, the linear approximation estimates $y(t) = 0.27$. This compares with the exact value of $0.52^2 = 0.2704$.

5.3 Other approaches to linearisation

Linearisation as described above consists of local approximation of a differentiable function by a linear function. Clearly, other types of approximation can be envisaged. For instance, a curve may be approximated by a finite number of straight line segments. Such an approximation can be undertaken graphically on an *ad hoc* basis whenever required.

Another approach to linearisation is to replace the function f, differentiable or not, by a linear multiplier k, chosen such that the error between $f(x)$ and kx is minimised when x moves over a range of values. Suppose that $x = a \sin \omega t$ is to be approximated, then k could be chosen to minimise R where

$$R = \frac{1}{2\pi} \int_0^{2\pi} [f(a \sin \omega t) - ka \sin \omega t]^2 \, d\omega t,$$

$$\frac{\partial R}{\partial k} = \frac{1}{2\pi} \int_0^{2\pi} [f(a \sin \omega t - ka \sin \omega t)] \, a \sin \omega t \, d\omega t.$$

Setting $\partial R/\partial k = 0$ yields

$$k = \frac{1}{\pi a} \int_0^{2\pi} f(a \sin \omega t) \sin \omega t \, d\omega t.$$

It can be seen that the best linear approximation involves finding the first term in a Fourier expansion representing a nonlinear waveform and neglecting the higher-order terms. This approach leads to the describing function method of analysis, previously discussed in Chapter 3.

Chapter 6

Determination of the qualitative behaviour of a nonlinear second-order system by linearisation (Lyapunov's first method)

6.0 Introduction

Whereas a linear system has only one type of behaviour everywhere in the phase plane, a nonlinear system may exhibit qualitatively different types of behaviour in different regions.

Let Σ be a nonlinear system having k qualitatively different types of behaviour in the phase plane. Suppose the system Σ to be linearised in each of the k regions, to yield linear approximations $\Sigma_1, \ldots, \Sigma_k$. It is a reasonable proposition that a knowledge of the behaviour of the system Σ in each of the k regions, can be obtained from a knowledge of the behaviour of each of the k linear approximations. The extent to which the proposition is valid was determined in a doctoral thesis in 1892 by A. M. Lyapunov. (Available in translation as Lyapunov (1966)). The approach outlined above and described below is sometimes called Lyapunov's first method.

6.1 Critical points of a system

Let Σ be a system described by a finite set of ordinary differential equations

$$\dot{x}_1 = f_i(x_1, \ldots, x_n), \quad i = 1, \ldots, n.$$

Then the *critical points* of the system are those points $p_j \, \varepsilon \, R^n$ for which $f_i(p_j) = 0$ for all $i \leqslant n$. At every critical point, all derivatives are zero and if the system is to come to rest with respect to the x variables, it can do so only at a critical point. However, a critical point may be one from which solutions emanate (called an unstable critical point or a *source*) or it may be one to which solutions move (called a stable critical point or *sink*). Critical points can be further sub-divided, according to the behaviour of solutions in their immediate vicinity. This point is taken up below for linear second-order systems.

6.2 Critical point analysis of a linear second-order system

Let Σ be a second-order linear system described by the equation

$$\dot{x} = Ax. \tag{6.1}$$

A critical point occurs at $x = \begin{pmatrix} 0 \\ 0 \end{pmatrix}$ and the nature of this critical point depends on the eigenvalues of the matrix A. Solutions of eqn. (6.1) either start or end at the critical point, or they encircle it. The critical point is named according to the behaviour of solutions around it as follows.

Behaviour of solutions	Name given to the critical point	Eigenvalues λ_1, λ_2	Behaviour in the phase plane
Solutions arrive monotonically at the critical point	*Stable node*	Both real, both negative	See Fig. 6.1
Solutions leave the critical point and grow monotonically	*Unstable node*	Both real, both positive	See Fig. 6.2
Only monotonic solutions, arriving at and leaving the critical point	*Saddle point*	Both real $\lambda_1 > 0$ $\lambda_2 < 0$	See Fig. 6.3
Solutions are oscillatory and arrive at the critical point	*Stable focus*	Both complex $\mathcal{R}(\lambda_1) = \mathcal{R}(\lambda_2) < 0$	See Fig. 6.4
Solutions are oscillatory and leave the critical point	*Unstable focus*	Both complex $\mathcal{R}(\lambda_1) = \mathcal{R}(\lambda_2) > 0$	See Fig. 6.5
Solutions are ellipses nested concentrically about the critical point	*Centre*	$\mathcal{R}(\lambda_1 = \mathcal{R}(\lambda_2) = 0$	See Fig. 6.6

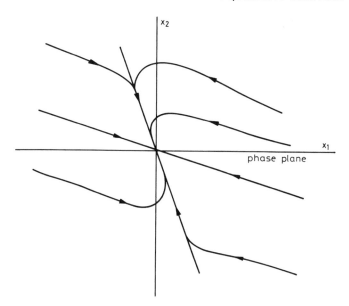

Fig. 6.1 *Phase portrait — a stable node*

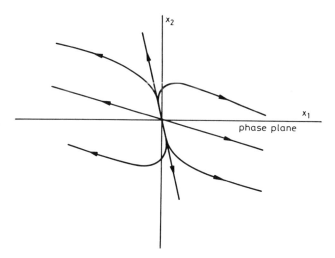

Fig. 6.2 *Phase portrait — an unstable node*

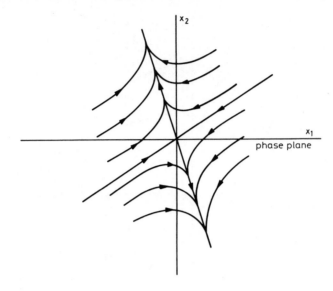

Fig. 6.3   *Phase portrait — a saddle-point*

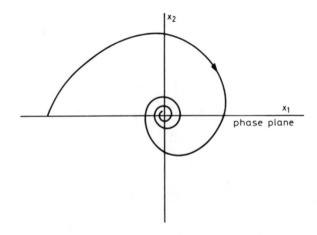

Fig. 6.4   *Phase portrait — a stable focus*

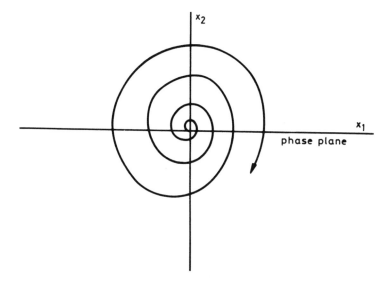

Fig. 6.5   *Phase portrait — an unstable focus*

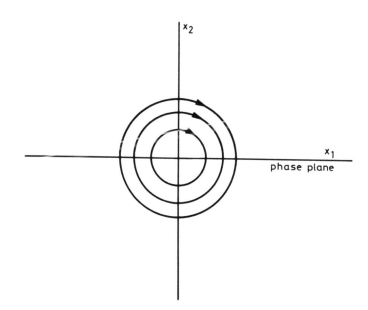

Fig. 6.6   *Phase portrait — a centre*

Figure 6.7 summarises the information in the table concisely, in terms of properties of the system matrix A.

Special cases not covered in the table are considered in Section 6.6.

The remainder of the table relates the eigenvalues to the type of critical point and gives a sketch of typical behaviour in the phase plane.

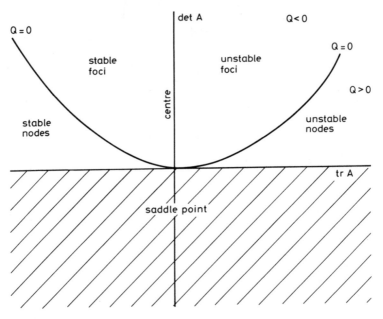

Fig. 6.7   *The nature of the critical points in terms of properties of the system matrix A. Note that tr A denotes the trace of the matrix A. Q is used to denote the discriminant in the characteristic equation for A*

Notice that the qualitative behaviour of the system over the entire phase plane is determined by the nature of the critical point at the origin. Linear systems have only one critical point and only one type of behaviour. In contrast, nonlinear systems may have multiple critical points and qualitatively different types of behaviour in different regions of the phase plane.

6.3 Critical points of nonlinear systems

The qualitative behaviour of a nonlinear system is largely governed by the locations and types of its critical points. The behaviour of the system between critical points can be determined approximately from a knowledge that the different types of behaviour around different critical points must be connected by smooth transitional behaviour.

6.4 Critical point analysis of a second-order nonlinear system

Consider a system described by the two first-order equations

$$\dot{x}_1 = f_1(x_1, x_2), \tag{6.2}$$
$$\dot{x}_2 = f_2(x_1, x_2).$$

The critical points $p \in \mathbb{R}^2$ are easily found by solving the equations $f_1 = f_2 = 0$.

The nature of the critical points p_j can often be found as follows. First the equations (6.2) are linearised to yield a matrix A whose ijth component is $\partial f_i / \partial x_j$. The A matrix is evaluated numerically at each of the critical points p_k, $k = 1, \ldots, r$ to obtain matrices A_k, $k = 1, \ldots, r$. The eigenvalues of each matrix A_k are determined and hence the nature of the critical point of each of the r linear systems is determined. The final postulate is that the nature of each critical point p_k of the nonlinear system can be inferred from the eigenvalues of the matrix A_k of the equivalent linear system.

The postulate is reasonable for any function f that can be represented as a Taylor series near to each critical point. In some region near to the critical point, the linear term is dominant and in the limit, at the critical point, the function and its linear approximation have the same behaviour.

Let the system be representable near the kth critical point by the equation

$$\dot{x} = A_k x + h(x),$$

then the postulate holds (i.e., the nature of the critical point p_k can be inferred from the eigenvalues of p_k) provided that:

(a) $\mathcal{R}(\lambda_1) = \mathcal{R}(\lambda_2) \neq 0$

(b) $\displaystyle \lim_{\xi \to 0} \frac{h(\xi)}{\|\xi\|^{1+p}} = 0$ for some $p > 0$.

These conditions are sometimes called Lyapunov's first theorem. A proof can be found in Driver (1977). Note carefully statement (a). It means that when the linearised system has a centre as its critical point, the condition will not necessarily carry over to the nonlinear system. The second statement provides a rigorous method by which to test the function f. It is clear (agreeing with intuition) that when f can be represented by a power series, that the statement (b) will be satisfied.

6.5 System behaviour near to a critical point

Consider again the linear second-order system described by eqn. (6.1). There are four basic types (node, saddle-point, focus, centre) of critical points.

Suppose that the matrix A has distinct eigenvalues λ_1, λ_2 with corresponding eigenvectors e_1, e_2 and modal matrix $E = (e_1|e_2)$. Then, putting $z = E^{-1}x$ diagonalises eqn. (6.1) to yield a solution

$$z(t) = e^{(E^{-1}AE)t} z(0),$$

or

$$\left.\begin{array}{l} z_1(t) = e^{\lambda_1 t} z_1(0), \\ z_2(t) = e^{\lambda_2 t} z_2(0), \end{array}\right\} \tag{6.3}$$

since

$$E^{-1}AE = \begin{pmatrix} \lambda_1 & 0 \\ 0 & \lambda_2 \end{pmatrix}.$$

Equation (6.3) can be represented in the phase-plane in z_1, z_2 axes that are coincident with the eigenvectors e_1, e_2. Alternatively, the solution can be represented in the original x_1, x_2 axes in which case the two fundamental modes in the solution (eqn. (6.3)) move along the eigenvectors. Taking the above discussion into account, it is clear how the system will behave at a node or at a saddle-point. At a node where $|\lambda_1| > |\lambda_2|$, the solution will move more rapidly along the e_1 direction than along the e_2 direction. The e_2 direction thus becomes an asymptote for the solutions. e_1 and e_2 are called *fast* and *slow eigenvectors* respectively. At a saddle-point where $\lambda_1 < 0$, $\lambda_2 > 0$, the component in the e_1 direction decays away, leaving only the unstable solution in the e_2 direction as an asymptote. Although the curvature of the solutions depends on the numerical values of λ_1 and λ_2, the form of the solutions is determined qualitatively by the above considerations. For the linear case the solutions are illustrated graphically in Fig. 6.8.

For the nonlinear case, the solution in the immediate vicinity of the critical point is the same as that for the linear system since here the linear term dominates. Further from the critical point, the nonlinearity modifies the behaviour. Clearly, a severe nonlinearity will modify the behaviour rapidly as the distance from the critical point increases. However, provided the nonlinearity is continuous, such modification will be smooth and progressive.

Around a focus, the trajectories of a linear system form logarithmic spirals in the phase plane. The rate of approach to or departure from the focus depends on the magnitude of the real part of the eigenvalues (recall that in this case they are a complex pair). For the nonlinear system, the behaviour in the immediate vicinity of the focus is as for the linear system, with increasing departure from the logarithmic spiral pattern as the distance from the critical point increases. Around a centre the trajectories of a linear system form nested ellipses.

The direction of rotation of trajectories around a focus or centre can be determined from a knowledge of the signs of x_1 and $\dot{x}_2$. For instance, let $(0, 0)$ be a focus in the phase plane and let ε be a small positive quantity. Then, if at the point $(0, \varepsilon)$, $\dot{x}_1 > 0$ and at $(0, -\varepsilon)$, $\dot{x}_1 < 0$, clockwise encirclement of the critical point by the trajectories is indicated.

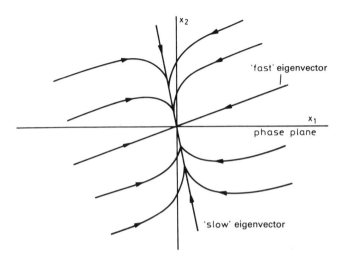

Fig. 6.8 *Fast and slow eigenvectors at a node*

6.6 Special cases — not covered in the table of Section 6.2

6.6.1 *Repeated eigenvalues $\lambda_1 = \lambda_2$*
All solutions are asymptotic to the single eigenvector *e* (Fig. 6.9).

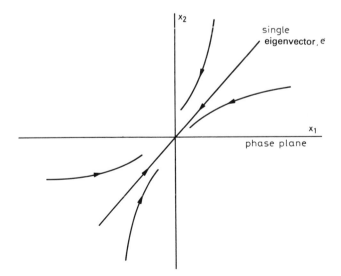

Fig. 6.9 *Phase portrait for a system with repeated eigenvalues*

6.6.2 Repeated eigenvalues — special case

If the system equations have the form

$$\dot{x}_1 = ax_1,$$

$$\dot{x}_2 = ax_2.$$

yielding the solution:

$$x_1(t) = c_1 e^{at}, \quad x_2(t) = c_2 e^{at}$$

then each trajectory has a constant slope c_2/c_1, determined by the initial conditions. The node is at the centre of a star of straight line trajectories (Fig. 6.10). (It will be seen that in this case, every vector in the plane is an eigenvector of the system.)

6.6.3 One eigenvalue is zero

Say $\lambda_1 = 0, \lambda_2 \neq 0$.

An example will be used to illustrate this case.

$$\left. \begin{array}{l} \dot{x}_1 = x_2, \\ \dot{x}_2 = -ax_2, \end{array} \right\} \lambda_1 = 0, \quad \lambda_2 = -a, \quad e_1 = \begin{pmatrix} 1 \\ 0 \end{pmatrix}, \quad e_2 = \begin{pmatrix} 1 \\ -a \end{pmatrix}.$$

Solutions are as shown in Fig. 6.11.

6.7 Examples

6.7.1 Example 1

$$\dot{x}_1 = x_2,$$

$$\dot{x}_2 = -x_1 - x_1^2 - x_2.$$

Critical points are at $(0, 0)$ and $(-1, 0)$. At $(0, 0)$ the equation can be rewritten

$$\begin{pmatrix} \dot{x}_1 \\ \dot{x}_2 \end{pmatrix} = \begin{pmatrix} 0 & 1 \\ -1 & -1 \end{pmatrix} \begin{pmatrix} x_1 \\ x_2 \end{pmatrix} + \begin{pmatrix} 0 \\ -x_1^2 \end{pmatrix}.$$

so that the A matrix of the linear approximation at $(0, 0)$ is given by

$$A = \begin{pmatrix} 0 & 1 \\ -1 & -1 \end{pmatrix}.$$

The equation has the form

$$\dot{x} = Ax + h(x),$$

and h is required to satisfy the condition

$$\lim_{\xi=0} \left| \frac{h(\xi)}{\|\xi\|^{1+p}} \right| = 0 \quad \text{for some } p > 0,$$

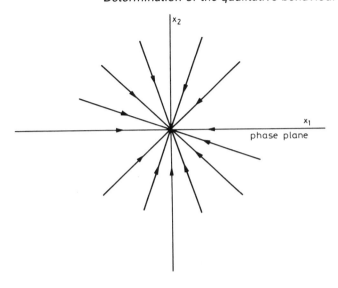

Fig. 6.10   *Phase portrait for a system with repeated eigenvalues — special case*

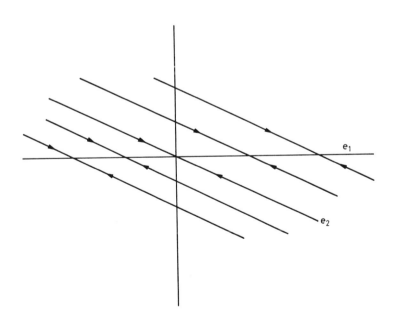

Fig. 6.11   *Phase portrait for a system having one zero eigenvalue*

in order that Lyapunov's first method shall be applicable.

$$\lim_{\xi \to 0} \frac{h(\xi)}{\|\xi\|^{1+p}} = \lim_{x \to 0} \frac{x_1^2}{\|x\|^{1+p}} \leq \lim_{x \to 0} \frac{x^2}{\|x\|^{1+p}}$$

$$= 0 \quad \text{for any } p \text{ in the interval } (0, 1).$$

The eigenvalues of the matrix A are $\lambda_{1,2} = -\frac{1}{2} \pm j\sqrt{3}/2$. Thus $(0, 0)$ is a stable focus when $x_1 = 0$ and $x_2 > 0$. Then, from the A matrix, $\dot{x}_1 > 0$. Thus the direction of rotation around the critical point $(0, 0)$ is clockwise.

At $(-1, 0)$,

$$A_{(-1,0)} = \begin{pmatrix} 0 & 1 \\ 1 & -1 \end{pmatrix}.$$

The nonlinear term still has the same form and still satisfies the required limit condition.

The eigenvalues of A are $\lambda_{1,2} = -\frac{1}{2} \pm \sqrt{5}/2$. Thus $(-1, 0)$ is a saddle-point, hence the eigenvalues are real and of opposite sign.

The eigenvectors of $A_{(-1,0)}$ are

$$e_1 = \begin{pmatrix} 1 \\ \dfrac{-1 + \sqrt{5}}{2} \end{pmatrix}, \quad e_2 = \begin{pmatrix} 1 \\ \dfrac{-1 - \sqrt{5}}{2} \end{pmatrix}.$$

The eigenvectors form fundamental solutions near to the point $(-1, 0)$.

Since all the eigenvalues of the two A matrices at $(0, 0)$ and $(-1, 0)$, respectively, have non-zero real part and since the nonlinearity obeys the laid-down limit test, the behaviour at the critical points of the original nonlinear system is the same as that found from the linear approximations. Transferring the information to a phase plane diagram yields Fig. 6.12.

By informed guesswork, the figure can be extended to give estimates of behaviour in other regions of the phase plane. Some guidelines are:

(*a*) Trajectories cannot cross except at critical points.
(*b*) The behaviour of a system with continuous equations will undergo a smooth transition from the locally linear behaviour around critical points. In particular the eigenvector directions, straight over the whole phase plane for a linear system, can be expected to curve as they move away from their critical point.
(*c*) Neighbouring trajectories will in general run in the same direction unless there is an underlying discontinuity present.

Using the rules, an estimated phase-portrait has been drawn for the system (Fig. 6.13). At this point it is of course always possible to compute a few trial trajectories numerically to confirm the form of the phase-portrait and to increase its accuracy.

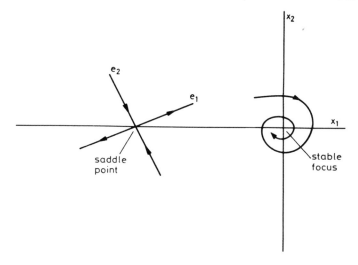

Fig. 6.12 *The nature of the critical points for Example 1*

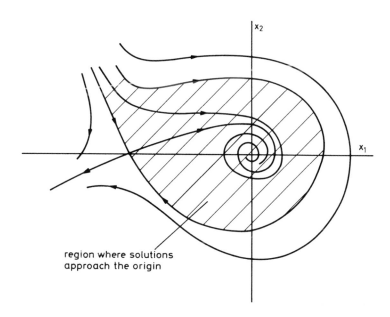

Fig. 6.13 *An estimated phase portrait for Example 1*

Notice in Fig. 6.13 that the trajectories that are extensions of the eigenvectors from $(-1, 0)$ separate the phase plane into regions. They are called *separatrices*.

Notice that it is also possible to estimate an area, shaded in Fig. 6.13, where solutions are asymptotically stable to the origin.

6.7.2 Example 2

$$\dot{x}_1 = -4x_1 + x_2 + 4x_1^2,$$

$$\dot{x}_2 = x_1 - 4x_2 - x_1^2.$$

Critical points are $(0, 0)$ and $(1, 0)$.
Linearising

$$A_{(0,0)} = \begin{pmatrix} -4 & 1 \\ 1 & 4 \end{pmatrix}, \qquad A_{(1,0)} = \begin{pmatrix} 4 & 1 \\ -1 & -4 \end{pmatrix}$$

At $(0, 0)$, $\lambda_1 = -5$, $\lambda_2 = -3$.
Eigenvectors are

$$e_1 = \begin{pmatrix} 1 \\ -1 \end{pmatrix}, \qquad e_2 = \begin{pmatrix} 1 \\ 1 \end{pmatrix}.$$

$(0,0)$ is a stable node. e_1 is the fast eigenvector and e_2 is the slow eigenvector (although the two time constants are of the same order).

At $(1,0)$, $\lambda_1 = \sqrt{15}$, $\lambda_2 = -\sqrt{15}$.
Eigenvectors are

$$e_1 = \begin{pmatrix} 1 \\ -0.127 \end{pmatrix}, \qquad e_2 = \begin{pmatrix} 1 \\ -7.87 \end{pmatrix}.$$

$(1, 0)$ is a saddle point.

The conditions for Lyapunov's first theorem to be applicable are satisfied. Hence the behaviour of the nonlinear system can be sketched in the phase plane (Fig. 6.14).

(See also Example 5 of Chapter 7 where Lyapunov's first method is applied to analyse the motion of a pendulum.)

6.8 Time-varying problems

The equation

$$\dot{x} = g(t, x),$$

where $x \in R^n$ and g is a given n-dimensional function, $g_i: R^{n+1} \to R^1$ is called a time-varying (sometimes non-autonomous) problem.

When eqn. (6.4) can be written in the so-called quasi-linear form

$$\dot{x} = A(t)x + f(t, x),\tag{6.5}$$

then the stability theorem below is applicable.

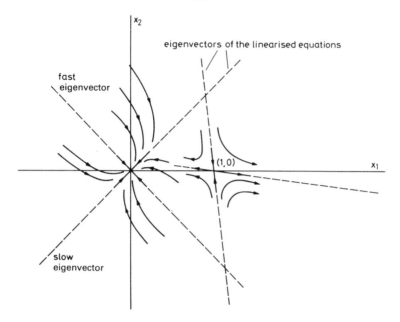

Fig. 6.14 *An estimated phase portrait for Example 2*

6.8.1 Theorem 6.1
Assume that in eqn. (6.5):

(a) $f(t, x)$ is continuous for $\|x\| < a$ and for $t \geqslant t_0$.
(b) A continuous non-negative function $\alpha(t)$ exists such that

$$\int_0^\infty \propto (t)\, dt < \infty \quad \text{and} \quad \|f(t, x)\| \leqslant \alpha(t)\|x\|.$$

(c) The equation $\dot{x} = A(t)x$ is uniformly stable.

Then the null solution of eqn. (6.5) is uniformly stable. The proof of the theorem can be found in Aggarwal (1974).

(Recall that the linear system $\dot{x} = A(t)x$ is *uniformly stable* if there exists a positive constant M such that

$$\|\Phi(t),\ \Phi^{-1}(\tau)\| < M \quad \text{for } t_0 \leqslant \tau \leqslant t < \infty,$$

for any fundamental matrix Φ of $A(t)$.)

6.9 Obtaining information on global stability properties from the linear approximation

(*Note*. This section requires knowledge of Lyapunov's second method which is the subject of Chapter 7.)

Consider the equation

$$\dot{x} = f(x), \quad x \in \mathbb{R}^n, \quad f(0) = 0. \tag{6.6}$$

6.9.1 Theorem 6.2

Denote by $J(x)$ the Jacobian matrix $\partial f/\partial x$. Solutions of eqn. (6.6) are globally asymptotically stable to the origin if there exists a positive definite symmetric matrix P such that

$$G = x^T(PJ(x) + J^T(x)P)x$$

is negative definite.

This can be proved by setting as a Lyapunov function

$$V(x) = x^T P x,$$

$$\dot{V}(x) = f^T(x)Px + x^T Pf(x),$$

but

$$f(x) = \int_0^1 J(\tau x)x \, d\tau.$$

Hence

$$\dot{V}(x) = x^T \left(\int_0^1 (J^T(\tau x)P + PJ(\tau x)) \, d\tau \right) x$$

and from the definition of G, $\dot{V}$ is negative definite and by Lyapunov's direct method, the system is globally asymptotically stable to the origin.

It also follows by putting $P = I$ (the identity matrix) that the system is globally asymptotically stable if the matrix $J(x) + J^T(x)$ has eigenvalues λ_i satisfying $\lambda_i < -k$ for all i, for some positive value of k.

The theorem can be reworked, see for instance Willems (1970), in terms of eqn. (6.6) written in the form

$$\dot{x} = A(x)x,$$

the result being obtained that the system is globally asymptotically stable to the origin provided that the eigenvalues of the matrix $A^T(x) + A(x)$ satisfy $\lambda_i < -k$, for all x, for some positive constant k.

Chapter 7

Lyapunov's second or direct method

7.0 Introduction

Let Σ be a system of any order, linear or nonlinear, for which the concept of energy is meaningful. Denote the total energy in the system by V. For instance, if Σ is a mechanical system then V is the sum of potential and kinetic energies. If Σ is an electrical system then V is the sum of electrical and magnetic energies. Clearly for a physically realisable system, V is a real number satisfying the condition $V \geq 0$. Suppose next that at every time t the condition $\dot{V}(t) < 0$ applies, then $V(t) \to 0$ as $t \to \infty$. The system will tend to a zero energy state and will be asymptotically stable. When Σ is represented by equations that cannot be solved by any analytical method, it may still be possible to evaluate the total energy V in the system and to calculate $\dot{V}$ and hence to determine stability.

Even in the case where the variables in the system Σ have no known physical significance, a mathematical stability test analogous to that outlined above is still possible.

The method is the most powerful available for nonlinear system analysis. With its aid, regions within which all solutions are attracted to a particular critical point, can be approximately delineated. The method is also useful in the synthesis of feedback loops of guaranteed asymptotic stability — particularly useful in the synthesis of adaptive systems.

An alternative geometric interpretation of the Lyapunov stability criterion is both useful and providing of insight.

Assume the origin of the phase plane to be surrounded by an arbitrary family of nested closed curves. If we can guarantee for every member of the family of closed curves, that solution trajectories always pass from outside to inside, and never the converse, then all solutions must approach the origin of the phase plane and asymptotic stability is ensured.

Supporting references for the chapter are Lyapunov (1966), La Salle and Lefschetz (1961) and Willems (1970).

7.1 Preliminary definitions

Let X be the n-dimensional space R^n. Let $\mathrm{x} \in X$ and let U be a region containing the origin of X. Let V be a function, $V: X \to \mathsf{R}^1$ (i.e., V is a scalar-valued function of a vector argument).

The function $V(x)$ is *positive definite* in U if for all x in U:

(a) V has continuous partial derivatives,
(b) $V(0) = 0$,
(c) $V(x) > 0$ if $x \neq 0$.

If (c) is replaced by (d), the function is *positive semi-definite*.

(d) $V(x) \geqslant 0$.

Negative definite and *negative semi-definite* functions are defined in an obvious way by reversal of the inequalities (c) and (d) respectively.

If in a region U containing the origin of $X = \mathsf{R}^n$, a function $V: X \to \mathsf{R}^1$ satisfies the condition $dV/dP > 0$, for every vector P radiating from the origin of X, then V is said to be a *strictly increasing function* on U.

Let U be a region containing the origin of $X = \mathsf{R}^n$. If a function $V: X \to \mathsf{R}^1$ satisfies the conditions:

(a) V is positive definite and strictly increasing on U;
(b) $\dot{V}$ is negative semi-definite on U.

Then V is called a *Lyapunov function* on U.

7.2 Lyapunov's second (or direct) stability theorem

Let Σ be a dynamic system with n-dimensional state space X. Let the origin in X be a critical point of X.

If on a region U, containing the origin of X, there exists a Lyapunov function for which $\dot{V} < 0$ on U, then the origin is a stable critical point and all solutions originating in the region U approach the origin asymptotically.

Note. By a change of axes, any critical point can be brought to the origin of phase space and the Lyapunov theorem applied.

7.3 Cetaev instability theorem

(A converse of the Lyapunov stability theorem.)

Define Σ, X, U as in the Lyapunov theorem above. If:

(a) V is positive definite on U,
(b) $\dot{V}$ is positive definite on U,

then the origin is an unstable critical point.

7.4 Geometric interpretation

As discussed in the introduction to this chapter, the Lyapunov second theorem can be considered from a geometric viewpoint.

The function $\dot{V} = dV(x)/dt$ can be expressed as the inner product

$$\dot{V} = \langle \nabla V, \dot{x} \rangle,$$

where, as usual, ∇V denotes the gradient vector of V.

Now turn to the phase portrait for the two-dimensional system Σ. The solution at any particular time t_1 is denoted by $x(t_1)$ and the tangent to this motion by the vector $\dot{x}(t_1)$. Consider the contour $V = k$ in the phase plane, for some value of k. The vector ∇V is an outward normal to the curve $V = k$. The projection of the vector $x(t_1)$ — representing the instantaneous motion — on to the outward normal to the contour $V = k$ is given by

$$\frac{\langle \nabla V, \dot{x} \rangle}{\|\nabla V\|} \frac{\nabla V}{\|\nabla V\|}.$$

Now if V is a Lyapunov function, the inner product in the above expression cannot be positive. Thus, the solution can have no component in the direction of the outward normal. It must move either along the $V = k$ contour or in a direction of decreasing V, having a component in the direction of the inward-pointing normal.

Fig. 7.1 may aid visualisation of the above concepts.

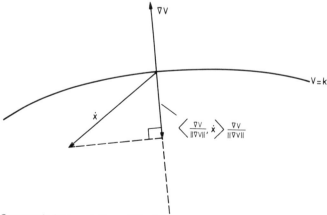

Fig. 7.1 *Geometric interpretation of the Lyapunov stability theorem*

7.5 Determination of the extent of a region of asymptotic stability around the origin

In those cases where a system has a finite region U that is asymptotically stable to the origin it will usually be the aim of Lyapunov analysis to discover, at least approximately, the extent of the region U.

Moving outwards into phase space from the origin, the limit of asymptotic stability may be set by $\dot{V}$ failing to remain negative definite. Calculation of a stability region may then proceed as will be seen in Example 3 of Section 7.7.

In other cases, the limit may be indicated by the function V failing to remain strictly increasing (this case will be seen in Example 4 of Section 7.7).

7.6 The search for a Lyapunov function

A major disadvantage in Lyapunov's direct method is that stability is assured 'if there exists a Lyapunov function'. However, it may not be clear where to begin in searching for a suitable function and if an arbitrary function is tried it will in general yield no useful information. Some suggested guidelines are:

(a) If the total energy in a system with physically meaningful variables can be quantified then this may form a suitable V function (Example 1 in the following section).

(b) If the system equations can be put in the form of Liénard's equation (Chapter 10) then, as is shown in that chapter, a Lyapunov function can always be constructed.

(c) If the system is a gradient system then, as shown in Chapter 11, a natural Lyapunov function exists.

(d) An ellipse of the form $V = ax_1^2 + bx_2^2$, with a and b to be determined, can be tried (Example 2).

(e) If the V function of (d) fails, the function $V = ax_1^2 + bx_2^2 + hxy$ may succeed (Example 3). In order that V shall remain positive definite, h must satisfy the condition $h^2 < ab$.

(f) From a detailed investigation of the equations or of the physics of the problem, a V function may suggest itself (Example 4).

(g) The system can always be decomposed as demonstrated in Chapter 11. It may then be analysed by a Lyapunov-like method.

7.7 Examples

7.7.1 Example 1
(A simple example with a known solution)
The linear electric circuit, Fig. 7.2, is described by the equations

$$L\frac{di}{dt} + iR + v_c = 0,$$

$$v_c = \frac{1}{C}\int i\,dt,$$

or

$$\frac{di}{dt} = \frac{1}{L}(- iR - v_c),$$

$$\frac{dv_c}{dt} = \frac{1}{C}i.$$

(7.1)

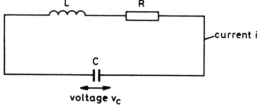

Fig. 7.2 *The electric circuit for Example 1*

Total energy V is known to be described by the equation

$$V = \tfrac{1}{2}Li^2 + \tfrac{1}{2}Cv_c^2.$$

This is a positive definite and strictly increasing function.

$$\dot{V} = Li\frac{di}{dt} + Cv_c\frac{dv_c}{dt}.$$

(7.2)

Substituting for di/dt and dv_c/dt from eqn. (7.1) yields

$$\dot{V} = Li\left(\frac{1}{L}(- iR - v_c)\right) + Cv_c\frac{1}{c}i = - i^2R.$$

(7.3)

Thus $\dot{V} < 0$ always, provided $i \neq 0$ and provided that R is positive.

(If $R = 0$, $\dot{V} \leq 0$, stability but not asymptotic stability is guaranteed. Under these conditions, the system is conservative and V remains constant.)

7.7.2 Example 2

Denote by V_l the function V evaluated along the trajectories of the linearised system. V_l is sometimes useful as an intermediate aid in setting up a Lyapunov function, as this example demonstrates.

$$\dot{x} = - x - 4y - x(x^2 + y^2),$$
$$\dot{y} = x - y + y(x^2 - y^2).$$

We try

$$V = ax^2 + by^2$$

as a Lyapunov function, with a, b to be determined.

$$\dot{V}_l = 2(ax\dot{x} + by\dot{y})$$
$$= 2(ax(- x - 4y) + by(x - y)).$$

Putting $a = 1$, $b = 4$, the terms in xy are eliminated and

$$\dot{V}_l = - 2x^2 - 8y^2.$$

For the full nonlinear system with these values of a and b,

$$\dot{V} = - 2x^2 - 8y^2 + 2x(- x(x^2 + y^2)) + 8y(y(x^2 - y))$$
$$= - 2x^2 - 8y^2 - 2x^4 - 8y^4 + 6x^2y^2.$$

To show that $\dot{V} < 0$ for $x, y = 0$, 'complete the square' to give

$$V = - 2x^2 - 8y^2 - (\sqrt{2}x^2 - \sqrt{8}y^2)^2 - 2x^2y^2 < 0, \quad x, y \neq 0.$$

Since V is positive definite and strictly increasing on the whole phase plane while $\dot{V}$ is negative definite, the system is globally, asymptotically stable to the origin.

7.7.3 Example 3

$$\left.\begin{array}{l} \dot{x} = - 4x + y + 4x^2, \\ \dot{y} = x - 4y - x^2. \end{array}\right\}$$

Try

$$V = ax^2 + by^2,$$
$$\dot{V} = 2ax\dot{x} + 2by\dot{y}.$$

Substitute for the system equations, linearised at the origin to give

$$\dot{V}_l = 2(ax(- 4x + y) + by(x - 4y))$$
$$= 2(- 4ax^2 + axy + bxy - 4by^2).$$

On this occasion, no choice of positive a, b can remove the terms in x, y. Accordingly we try

$$V = ax^2 + by^2 + 2hxy,$$

which is positive definite and strictly increasing provided that a, $b > 0$ and $h^2 < ab$.

$$\dot{V}_l = 2(- 4ax^2 + (a + b)xy - 4by^2) + 2h(x\dot{y} + \dot{x}y)$$
$$= 2(-4ax^2 + (a + b)xy - 4by^2) + 2h(x^2 - 4xy - 4xy + y^2).$$

If we set $a = b = 4$ and $h = 1$ then the terms in xy are eliminated and $\dot{V}_l < 0$ (notice, however, that other choices of a, b, h could also achieve this objective).

We now determine $\dot{V}$ for the full system using the values $a = b = 4, h = 1$.

$$\dot{V} = (8x + 2y)(- 4x + y + 4x^2) + (2x + 8y)(x - 4y - x^2)$$
$$= - 30x^2(1 - x) - 30y^2.$$

Thus we have $x < 1 \to \dot{V} < 0$.

A little manipulation will show that

$$V < 4 \to x < 1 \to \dot{V} < 0.$$

Thus, the contour $V = 4$ in the phase plane defines a region of asymptotic stability to the origin. The equation of the contour is

$$4x^2 + 4y^2 + 2xy = 4.$$

The region of asymptotic stability defined by the contour is sketched in Fig. 7.3.

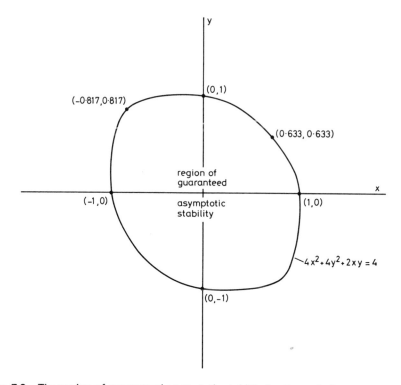

Fig. 7.3 *The region of guaranteed asymptotic stability for Example 3*

7.7.4 Example 4

$$\left. \begin{array}{l} \dot{x} = y, \\ \dot{y} = -x - x^2 - y. \end{array} \right\}$$

Try

$$V = ax^2 + by^2,$$
$$\dot{V} = 2(axy + by(-x - x^2 - y)).$$

Here we require $a = b = 1$ to eliminate the terms in xy but a term in x^2y remains. To proceed, we could introduce a term $2x^3/3$ into the express for V, i.e., put $V = x^2 + y^2 +$ then $2x^3/3$ $\dot{V} = -2y^2$.

The limit of stability around the origin will be set by the V function failing to be strictly increasing. V takes on a maximum value of $V = \frac{1}{3}$ on the x-axis at the point $(-1, 0)$. V is therefore a Lyapunov function within the region bounded by the curve

$$x^2 + y^2 + 2x^3/3 = \frac{1}{3}$$

and the region U is a region of asymptotic stability to the origin since $\dot{V}$ is negative definite.

It is instructive to analyse the problem by Lyapunov's first method. The point $(-1, 0)$ will be found to be a saddle-point located at the boundary of the stability region found by the second method.

7.7.5 Example 5

$$\ddot{y} + F\dot{y} + \sin y = 0, \quad 0 < F < 2.$$

The equation, which describes a pendulum, is analysed by both the first and second Lyapunov methods.

Analysis by the first method
Let $x_1 = y$, $x_2 = \dot{x}$, yielding

$$\dot{x}_1 = x_2,$$
$$\dot{x}_2 = -\sin x_1 - Fx_2.$$

Critical points occur at $(n\pi, 0)$, $n = 0, \pm 1, \pm 2, \ldots$.

When n is even the A matrix of the linear approximation is

$$A = \begin{pmatrix} 0 & 1 \\ -1 & -F \end{pmatrix},$$

$$\lambda_{1,2} = -\frac{F}{2} \pm \sqrt{\frac{F^2}{4} - 1}.$$

The critical points for n even are stable foci.

When n is odd, the A matrix of the linear approximation is

$$A = \begin{pmatrix} 0 & 1 \\ 1 & -F \end{pmatrix},$$

$$\lambda_{1,2} = -\frac{F}{2} \pm \sqrt{\frac{F^2}{4} + 1}.$$

The critical points for n odd are saddle points.

Analysis by the second method
Consistent with the total energy stored in a pendulum, we define

$$V = \frac{x_2^2}{2} + \int_0^{x_1} \sin z \, dz = \frac{x_2^2}{2} - \cos x_1 + 1,$$

$$\dot{V} = x_2 \dot{x}_2 + x_2 \sin x_1 = -Fx_2^2.$$

Along the x_1 axis, V is strictly increasing, provided that $-\pi < x < \pi$. A line of constant V through the points $(-\pi, 0)$, $(\pi, 0)$ defines a region of asymptotic stability to the origin.

(It is significant that the points $(-\pi, 0$ and $(\pi, 0)$ are the saddle-points found by the first method.)

The curve enclosing the region of asymptotic stability has the equation

$$x_2 = (2(1 + \cos x_1))^{1/2}.$$

By symmetry, it can be seen that each focus along the x_1-axis can be surrounded by a similar stability region (Fig. 7.4).

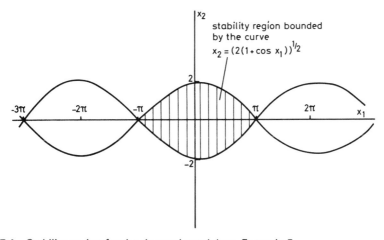

Fig. 7.4 *Stability region for the damped pendulum, Example 5*

Notice, finally, how complete a knowledge of the overall behaviour of the pendulum is obtained from nonlinear analysis.

7.8 Lyapunov's direct method for time-varying problems

Given the equation

$$\dot{x} = f(t, x), \quad x \in \mathbf{R}^n.$$

Assumptions:

(a) $f(t, x)$ is defined for all values of x and solutions exist for any choice of x_0.
(b) $f(t,x)$ is defined and continuous for $t > r$ for some given r.
(c) The equation has a unique solution for every x_0 and for every $t_0 > r$.
(d) $f(t, 0) = 0$ for all t.

7.8.1 Stability theorem
If a positive definite function $V(t, x)$ exists for which $\dot{V}(t, x)$ is negative definite then the solution $x(t) \equiv 0$ is asymptotically stable.

($V(t, x)$ is defined to be positive definite if there exists a continuous strictly increasing function ϕ for which $\phi(0) = 0$ and $V(t, x) > \phi(\|x\|)$ for $t > r$ and for all x in $\mathbf{R}^n$).

7.9 Use of the Lyapunov direct method as a design tool

It is sometimes useful to be able to synthesise a control loop of guaranteed stability. This is particularly so when adding an additional adaptive loop to an existing closed-loop control system. Kalman and Bertram (1960) have considered in detail how the Lyapunov direct method can be used as a design tool. Leigh and Li (1981) describes an application to ship steering in which the adaptive algorithm described is synthesised so as to be of guaranteed stability.

Here only the simplest possible example is given to illustrate the principle.

$$\dot{x}_1 = x_2,$$

$$\dot{x}_2 = -x_1 - x_2 + ku,$$

$$u = -x_1.$$

The equations describe a feedback control system. The gain k is to be chosen to ensure stability.

Take as a Lyapunov function

$$V = ax_1^2 + bx_2^2,$$

$$\dot{V} = -2bx_2^2 + 2(a - b - bk)x_1x_2.$$

To eliminate the term in x_1x_2, k must be chosen to satisfy the relation

$$k = \frac{a}{b} - 1$$

According to the choice of a and b in the equation for V, values for k in the range $(-1, \infty)$ will be found.

In this example a best possible result will be obtained by putting $a \gg b$.

7.10 Concluding remarks

When stability is guaranteed by the second method of Lyapunov, the system response may still be unsatisfactory in that it is highly oscillatory and/or takes an inordinately long time to settle down. For instance, a numerical iterative loop whose convergence was guaranteed by the second method, converged so slowly as to be quite useless in practice. This aspect is particularly important when the Lyapunov method is used as a design tool. The designer must ensure that a satisfactory rate of convergence is obtained.

Chapter 8

Envelope methods — the Popov and circle criteria for graphical analysis of a single feedback loop containing a nonlinear element

8.0 Introduction

This chapter is concerned with stability analysis of the loop shown in Fig. 8.1. The two methods to be described enclose the nonlinearity in a linear envelope. The linear envelope rather than the particular nonlinearity is then used in the subsequent analysis. Such an approach naturally leads to sufficient but not necessary stability conditions.

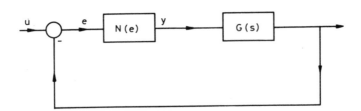

Fig. 8.1 *The feedback loop to be analysed by the Popov or circle method*

8.1 The Aizerman and Kalman conjectures

Before proceeding to describe graphical techniques for the analysis of a feedback loop containing a nonlinearity, it is instructive to consider two celebrated conjectures:

8.1.1 Aizerman's conjecture
The system of Fig. 8.1 will be stable provided that the linear system of Fig. 8.2 is stable for all values of k in the interval $[k_1, k_2]$ where k_1, k_2 are defined by the relation $k_1 \leq N(e)/e \leq k_2$ for all $e \neq 0$. (k_1, k_2 represent a linear envelope surrounding the nonlinearity.)

Aizerman's conjecture is false as has been shown by counter-example (Willems, 1970).

8.1.2 Kalman's conjecture

The system of Fig. 8.1 will be stable provided that the linear system of Fig. 8.2 is stable for all k in the interval $[\hat{k}_1, \hat{k}_2]$ where

$$\hat{k}_1 \leq \frac{dN(e)}{de} \leq \hat{k}_2$$

and where

$$k_1 \leq \frac{N(e)}{e} \leq k_2$$

and

$$\hat{k}_1 \leq k_1 \leq k_2 \leq \hat{k}_2.$$

Kalman's conjecture imposes additional requirements on the nonlinear characteristic but nevertheless it is also false — again shown by counter-example.

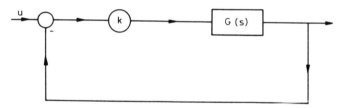

Fig. 8.2 *The linear feedback loop that is the subject of Aizerman's and Kalman's conjectures*

The failure of the two conjectures shows that feasible intuitive reasoning cannot be relied on in nonlinear situations. One reason for the failure of the conjectures is that instabilities may arise in a nonlinear system due to the effects of harmonics. These are, of course, entirely absent in linear systems.

8.2 The Popov stability criterion

Popov (1961) developed a graphical Nyquist-like criterion to examine the stability of the loop shown in Fig. 8.1.

$G(s)$ is assumed to be a stable causal transfer function. The input u must be bounded, continuous and must satisfy the condition

$$\int_0^\infty |u(t)|^2 \, dt < \infty.$$

The nonlinearity $N(e)$ must be a time-invariant and piecewise continuous function of e. $dN(e)/de$ must be bounded and $N(e)$ must satisfy the condition $0 < N(e)/e < k$ for some positive constant k. (Graphically, the last condition means that the curve representing N must lie within a particular linear envelope.)

A sufficient condition for global asymptotic stability of the feedback loop may then be stated:

If there exists any real number q and an arbitrarily small number $\delta > 0$ such that

$$R\{(1 + j\omega q)\, G(j\omega)\} + \frac{1}{k} \geq \delta > 0 \tag{8.1}$$

for all ω then for any initial state the system output tends to zero as $t \to \infty$.

(The proof can be found in Desoer (1965). It makes use of Lyapunov's direct method.)

To carry out a graphical test based on eqn. (8.1), a modified transfer function $G^*(j\omega)$ is defined by

$$G^*(j\omega) = R\{G(j\omega)\} + j\omega I\{G(j\omega)\} \equiv X(j\omega) + jY(j\omega). \tag{8.2}$$

Equation (8.2) becomes, in terms of X and Y

$$X(j\omega) - qY(j\omega) + \frac{1}{k} \geq \delta > 0.$$

The $G^*(j\omega)$ curve (the so-called Popov locus) is plotted in the complex plane. The system is then stable if some straight line, at an arbitrary slope $1/q$, and passing through the $-1/k$ point avoids intersecting the $G^*(j\omega)$ locus. Figs. 8.3 and 8.4 show two possible graphical results for stable and not necessarily stable situations respectively. (Recall that the test gives a sufficient condition for stability and that the feedback loop whose result is given in Fig. 8.4 is not necessarily unstable.)

8.3 The circle method

The circle method of stability analysis can be considered as a generalisation of Popov's method. Compared with that method it has two important advantages:

(a) It allows $G(s)$ to be open loop unstable;
(b) It allows the nonlinearity to be time varying.

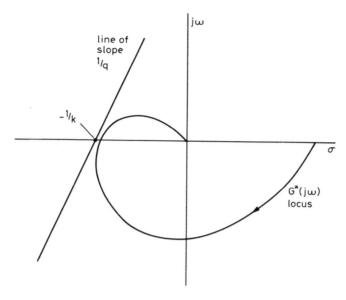

Fig. 8.3   *The Popov stability test — the control loop is guaranteed stable*

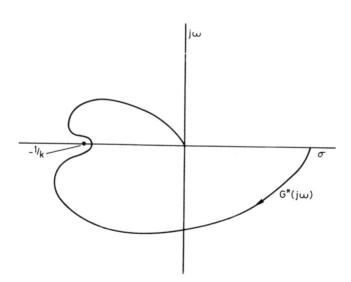

Fig. 8.4   *The Popov stability test. No line through the — 1/k point avoids intersection with the G*(jω) locus and the loop may be unstable*

The nonlinearity N is considered to lie within an envelope such that (Fig. 8.5).

$$Ae < N(e, t) < Be.$$

Then it is a sufficient condition for asymptotic stability that the Nyquist plot $G(j\omega)$ lies outside a circle in the complex plane that crosses the real axis at the points $-1/A$ and $-1/B$ and has its centre at the point

$$\tfrac{1}{2}\left[-\left(\frac{1}{A}+\frac{1}{B}\right)+\tfrac{1}{2}j\omega q\left(\frac{1}{A}-\frac{1}{B}\right)\right]$$

for some real value of q. (Here it is assumed that $1/A > 1/B$.)

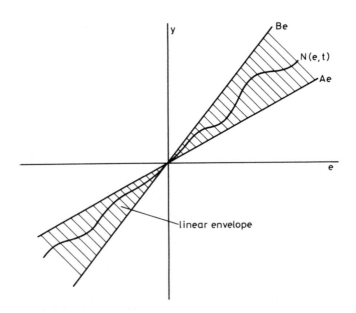

Fig. 8.5   *The linear envelope for the circle criterion*

This is the so-called generalised circle criterion. Notice that the centre of the circle depends on both frequency and choice of the value of q.

In return for a loss of sharpness in the result (remembering that the method gives a sufficient condition), q can be set equal to zero and then a single frequency invariant circle results (Fig. 8.6). The circle can be considered as the generalisation of the $(-1, 0)$ point in the Nyquist test for linear systems.

Further discussion and examples of the circle method can be found in Shinners (1972). The proof of the stability criterion can be found in the original paper, Zames (1966) or in Willems (1970). Willems' proof uses path integral representation of a quadratic form in conjunction with Lyapunov's direct method.

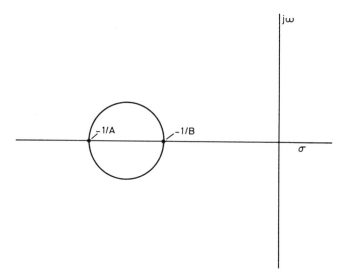

Fig. 8.6 *The circle criterion — the frequency independent circle that results if q is set equal to zero*

Chapter 9

Limit cycles and relaxation oscillations

9.0 Introduction

Oscillations were first studied as part of nonlinear mechanics and a long standing and comprehensive literature exists. In particular, there are fundamental theories concerning the existence of stable periodic solutions. Some knowledge of this mathematical background is helpful in interpreting the results of nonlinear system analyses.

Conversely, if it is required to construct a mathematical model of a stable periodic system, such as a biological oscillator, it will be essential to postulate a mechanism that has the necessary properties.

9.1 Limit cycle — definition

Let Σ be a second-order system and let C be a closed curve in the phase plane representation of Σ. Let x_0 be any point in the neighbourhood of C. If every trajectory $\pi(x_0, t)$ moves asymptotically onto C as $t \to \infty$ then C is called a *stable limit cycle* (Fig. 9.1). If every trajectory $\pi(x_0, t)$ moves asymptotically onto C as $t \to -\infty$ then C is called an *unstable limit cycle* (Fig. 9.2).

Limit cycles are essentially nonlinear phenomena. Notice from the definition that they are isolated — they are not part of a continuous family of closed curves — compare with the solution of the linear equation $\ddot{x} + x = 0$.

Notice that an unstable limit cycle, Fig. 9.2 can form the boundary of the region whose solutions are asymptotically stable to the origin.

Example of a limit cycle

The equations (Sanchez, 1968)

$$\dot{x}_1 = x_2 + x_1(1 - x_1^2 - y_1^2),$$

$$\dot{x}_2 = x_1 + x_2(1 - x_1^2 - y_1^2),$$

can be solved explicitly, and in polar coordinates the solution is

$$r(t) = (1 + \mu e^{-2t})^{1/2}, \quad \theta(t) = \phi - t,$$

where μ and ϕ depend on initial conditions.

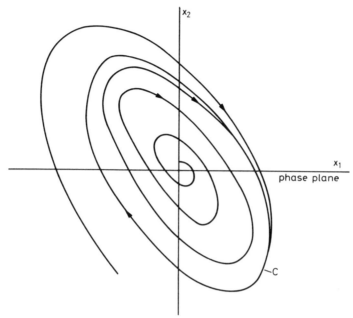

Fig. 9.1 *A stable limit cycle, C, in the phase plane*

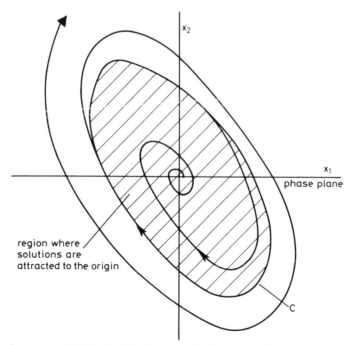

Fig. 9.2 *An unstable limit cycle, C, whose interior is a region of asymptotic stability to the origin*

The solution tends to the unit circle from all regions of the phase plane. The unit circle is a stable limit cycle for this equation.

9.2 Asymptotic stability

Let C be a closed nontrivial trajectory representing a periodic solution of the system Σ. Let $x(t)$ be any solution of the system in the neighbourhood of C.
Then if

$$\lim_{t \to \infty} d(C, x(t)) = 0,$$

where

$$d = \min \{\|y - x(t)\|, y \in C\},$$

then C is defined to be *orbitally asymptotically stable*.

9.2.1 Theorem 9.1
Let C be a closed nontrivial trajectory of the system Σ in R^n. Let p be any point in C. Let A be the linear approximation to Σ at p. Suppose that $n - 1$ eigenvalues of A satisfy $|\lambda| < 1$ then C is orbitally asymptotically stable (one of the eigenvalues will always be unity).

Hirsch and Smale (1974) prove the theorem and show that the result is independent of the choice of the point p.

It may further be required that when a periodic motion around a closed trajectory C is disturbed, not only does the motion return to the original trajectory but that when it does so it regains its original phase relationship. Such a closed trajectory C is then called a *periodic attractor*. It possesses a stronger property than asymptotic stability (for further details see Hirsch and Smale, 1974).

9.3 The Poincaré index and its implications in the phase plane

Let $\tilde{x}$ be a critical point in a planar vector field and let $\tilde{x}$ be enclosed by a closed curve S in the plane. Let R be a point on S that traverses the curve S in the clockwise direction. Let r be a vector that is based at R and which is at all times parallel to the vector field. The *Poincaré index* P is then defined as the number of times that r rotates as R traverses S (clockwise rotation being defined as positive). The following theorems have been put forward by Poincaré (1928).

(*a*) The index of a saddle-point is -1 and of any other critical point is $+1$.
(*b*) The index of a closed curve not surrounding a critical point is zero.
(*c*) The index of a closed curve containing one or more critical points is equal to the algebraic sum of their indices.

(*d*) The index of a closed curve that is a trajectory is always $+1$.
(*e*) Let S be a closed curve with respect to which the field vectors are directed either inwards or outwards at all points then $P = +1$.

Consequences are:

(*e*) A closed trajectory must contain at least one critical point of index $+1$ or several critical points, the algebraic sum of whose indices is $+1$.

9.3.1 Bendixson's first theorem

Let Σ be a second-order system whose flow can be represented by the equations

$$\left. \begin{aligned} \dot{x}_1 &= Q_1(x_1, x_2) \\ \dot{x}_2 &= Q_2(x_1, x_2) \end{aligned} \right\}$$

If the expression $(\partial Q_1/\partial x_1) + (\partial Q_2/\partial x_2)$ does not change sign in some region D of the phase plane then no periodic motion can exist in D.

Proof: Let S be a closed trajectory in a region D of the phase plane and assume that the expression changes sign. Then by Gauss' law

$$\oint_s (Q_1 \, dx_1 - Q_2 \, dx_2) = \iint \left(\frac{\partial Q_1}{\partial x_1} + \frac{\partial Q_2}{\partial x_2} \right) dx_1 \, dx_2 \triangleq F.$$

Now assume that S is a trajectory then

$$F = \oint (\dot{x}_1\dot{x}_2 \, dt - \dot{x}_2\dot{x}_1 \, dt) = 0,$$

so $(\partial Q_1/\partial x_1) + (\partial Q_2/\partial d_2)$ is either zero everywhere (which is not possible) or it changes sign in D, which is a contradiction.

9.3.2 Implications

(*a*) If a system has just one critical point whose index is not $+1$ then periodic motions are impossible.
(*b*) If a system has several critical points the algebraic sum of whose indices is not $+1$, periodic motions on closed curves enclosing all these critical points are impossible.
(*c*) If a system has one critical point with index $+1$, which is approached by trajectories as $t \to \infty$, periodic motions are impossible.

9.3.3 Bendixson's second theorem

Let a half trajectory $\pi^+(x, t)$, $t \geq t_0$ remain inside a finite region D in the phase plane without approaching any critical point then either the trajectory is a closed trajectory or it approaches asymptotically a closed trajectory. (The theorem can be applied to Van der Pol equations by constructing an annulus containing no critical point with trajectories entering the annulus at all points but with no trajectory leaving.)

Proof: See Lefschetz (1977).

9.3.4 Theorem 9.2
Let Σ be a second-order conservative system that can be described by the equations

$$\left.\begin{aligned} \dot{x}_1 &= x_2, \\ \dot{x}_2 &= f(x_1). \end{aligned}\right\}$$

Then system trajectories cross the x_1 axis vertically and cross the curve $f(x) = 0$ horizontally, except where $x_2 = 0$.

Proof: $dx_2/dx_1 = f(x_1)/x_2$ and the result is obvious.
(For this type of system it is instructive to plot $\int_0^{x_1} f(s)\, ds$ against x_1. There are theorems by Lagrange and Lyapunov relating to such graphs (Minorsky, 1947).

9.4 Relaxation oscillations

Relaxation oscillations are self-excited oscillations exhibiting discontinuous or quasi-discontinuous features. In a typical case, some physical quantity is able to exist at two different discrete magnitudes and it passes between the two levels very quickly. In a mathematical representation, the transition between the levels will be achieved in zero time. The instantaneous changes occur when the system equations become degenerate and cease to describe the system motion.
A very simple example is given below.

9.4.1 Example
Consider the circuit shown in Fig. 9.3 where for convenience the capacitor has unit capacitance and where the voltage across the non linear resistor is given by the graph shown alongside the circuit. We have

$$\frac{1}{c}\int i\, dt = -f(i), \quad i = -f'(i)\,\frac{di}{dt} \quad,$$

or

$$\frac{di}{dt} = \frac{-i}{f(i)}, \quad \text{where} \quad f'(i) = \frac{df(i)}{di}.$$

This equation is singular when $f'(i) = 0$ and the behaviour of the system is not completely determined by the equation. In fact, it is known by experiment that the system jumps from a singularity instantaneously as shown by the dotted arrows in Fig. 9.4. Desoer has suggested that the analysis of this type of problem can be undertaken rigorously as follows.

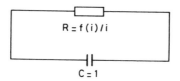

Fig. 9.3 *Electric circuit with nonlinear resistor*

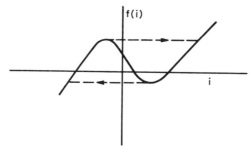

Fig. 9.4 *The characteristic of the nonlinear resistor with (superimposed in dotted lines) the path of the relaxation oscillation*

Introduce a small fictitious inductor ϵL in series in the circuit (every circuit must have some inductance). The resulting differential equations are then

$$\left.\begin{aligned}
\epsilon L \,\frac{di}{dt} &= -v_c - f(i), \\[2mm]
\frac{dv_c}{dt} &= -i,
\end{aligned}\right\}$$

(where v_c is the voltage across the capacitor).

These equations are no longer singular but the previous singular equation is obtained in the limit as $\epsilon \to 0$.

In summary, to analyse the relaxation oscillator it is necessary to plot the nonlinear characteristic in the phase plane and to supplement information from the differential equation with knowledge of the instantaneous jump phenomenon.

Chapter 10

Liénard's equation

10.0 Introduction

Certain nonlinear equations have been extensively studied and the solutions and stability properties are well understood.

One such equation is Liénard's equation. It is studied here because it is able to represent a wide class of simple but important nonlinear physical problems.

10.1 Liénard's equation — definition

The second-order scalar equation

$$\ddot{x} + f(x)\dot{x} + g(x) = 0, \tag{10.1}$$

with some restriction on the functions f and g, is known as Liénard's equation. It can be considered as a generalised mechanical system with nonlinear spring characteristic and nonlinear damping term. It can also represent a variety of nonlinear electric circuits.

If a problem can be put into the Liénard form, the stability properties and any limit cycle behaviour can be determined from a knowledge of the functions f and g.

10.1.1 Restrictions on the functions f and g
The function f is to be an even function of x.

The function g is to satisfy $g(0) = 0$ and to be monotone increasing.
Define

$$F(x) = \int_0^x f(s)\ ds, \quad G(x) = \int_0^x g(s)\ ds.$$

Then eqn. (10.1) can be written, introducing a new variable y,

$$\left.\begin{aligned} \dot{x} &= y - F(x), \\ \dot{y} &= -g(x). \end{aligned}\right\} \tag{10.2}$$

At equilibrium points, $y = F(x)$, $g(x) = 0$. Hence $x = 0$, $y = 0$ is the only critical point.

10.2 Analysis of Liénard's equation by Lyapunov's direct method

Referring to eqn. (10.2) define a Lyapunov function

$$V = y^2/2 + G(x),$$
$$\dot{V} = y\dot{y} + g(x)\dot{x} = -g(x)F(x).$$

The stability of the system can be tested directly using the standard Lyapunov function as above. In case a stability limit is to be estimated, it is useful to find constants a, l such that:

(a) $|x| < a \rightarrow \dot{V} < 0$;
(b) $G(x) < l \rightarrow |x| < a$.

then

$$V < l \rightarrow |x| < a \rightarrow \dot{V} < 0.$$

Thus $V < l$ defines a region of guaranteed asymptotic stability in the phase plane.

10.3 Examples

10.3.1 Example 1

$$\ddot{x} + \dot{x} + x^3 = 0.$$

This is a simple example of Liénard's equation with, $f(x) = 1$, $g(x) = x^3$. As in the standard solution we put

$$\dot{x} = y - F(x),$$
$$y = -g(x),$$
$$F(x) = \int_0^x f(s)\,ds = x,$$
$$G(x) = \int_0^x g(s)\,ds = x^4/4,$$
$$V = y^2/2 + G(x) > 0, \quad \dot{V} = -g(x)F(x) = -x^4 < 0.$$

The solution is globally asymptotically stable.

10.3.2 Example 2. Van der Pol's equation

$$\left. \begin{array}{l} \dot{x} = y - \epsilon(x^3/3 - x), \\ \dot{y} = -x. \end{array} \right\}$$

This is an example of Liénards equation with

$$f(x) = \epsilon(x^2 - 1), \quad g(x) = x,$$

$$F(x) = \epsilon(x^3/3 - x), \quad G(x) = x^2/2,$$

$$V = y^2/2 + G(x) = \frac{x^2 + y^2}{2},$$

$$\dot{V} = -g(x)F(x) = -\epsilon x \left(\frac{x^3}{3} - x\right)$$

$$= -\epsilon x^2 \left(\frac{x^2}{3} - 1\right).$$

Assume $\epsilon < 0$, then $x^2 < 3 \rightarrow \dot{V} < 0$.

We use the rule — find constants a, l such that:

(a) $|x| < a \rightarrow \dot{V} < 0$, $(a = \sqrt{3})$;

(b) $G(x) < l \rightarrow |x| < a$ $(l = 3/2)$.

Then $V < l$ implies asymptotic stability to the origin, i.e., the stability region is contained inside the closed curve.

$$V = \tfrac{3}{2} \quad \text{or} \quad x^2 + y^2 = 3.$$

In fact the Van der Pol equation has a limit cycle as a closed trajectory encircling the origin. In the case that we are considering ($\epsilon < 0$), the limit cycle is unstable and represents the true boundary of the region of asymptotic stability around the origin.

The result we have just obtained tells that at no point does the limit cycle come nearer the origin than the distance $\sqrt{3}$.

Now consider the effect of changing the sign of ϵ in the original equation. It can be seen that this is the same as reversing the direction of time — it results in the arrows in the phase portrait becoming reversed but a limit cycle remains a limit cycle. Now we have $V > 0$ and $\dot{V} > 0$ in a region around the origin satisfying $V < 3$ and by Cetaev's theorem this is an unstable region of solutions moving away from the origin. The solutions will tend to the unique closed trajectory (limit cycle) situated at least a distance $\sqrt{3}$ from the origin.

10.4 Existence of a limit cycle for Liénard's equation

In eqn. (10.1) assume that:

(a) f is an even continuous function satisfying $f(0) < 0$;

(b) g is an odd continuous function satisfying $xg(x) > 0$ for all $x \neq 0$ and g is Lipschitzian for all x;

(c) $F = \int_0^x f(s)\, ds \to \pm\,\infty$ as $x \to \pm\,\infty$;

(d) F has a single positive zero at $x = a$ and is monotone increasing for $x \geqslant a$.

Then eqn. (10.1) possesses a unique stable limit cycle.

Proof: Lefschetz (1977, p. 268) shows that these are sufficient but not necessary conditions. The conditions (a) to (d) above may appear complex at first sight but roughly all that they imply is that:

(a) $F(x)$ has the form shown in Fig. 10.1. Notice that there is negative damping for small x and positive damping for large x.

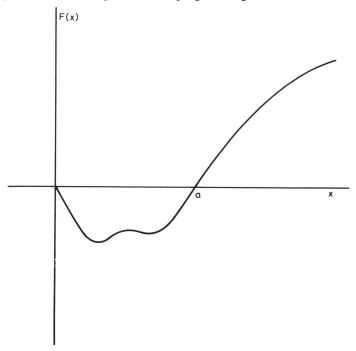

Fig. 10.1 *The required form of F(x) to satisfy the conditions of Section 10.4 (notice that the function need not be monotonic for x < a)*

(b) g, although nonlinear, has its characteristic in the first and third quadrants, passing through the origin.

10.5 A particular example of Liénard's equation that has a stable limit cycle

Consider the Van der Pol equation

$$\ddot{x} + 2\zeta(1 - x^2)\dot{x} + x = 0,$$

where ζ is a constant.

For $\zeta > 0$ it is known that the equation describes a stable oscillator. The equation can be put in Liénard's form by setting

$$f(x) = - 2\zeta(1 - x^2),$$
$$g(x) = x,$$

$$F(x) = \int_0^x - 2\zeta(1 - x^2) \, dx = - 2\zeta x + \frac{2\zeta x^3}{3},$$

$$G(x) = \frac{x^2}{2}.$$

conditions (*a*) to (*d*) above are satisfied and without further calculation it is confirmed that the equation possesses a unique stable limit cycle.

When the parameter ζ is small, say $\zeta < 0\cdot05$, the oscillation is nearly sinusoidal. As ζ increases, the behaviour changes to that of a relaxation oscillator; $\zeta = 10$ gives a markedly discontinuous behaviour.

Chapter 11

Gradient systems and system decomposition

11.0 Introduction

Gradient systems have some particularly simple properties. They cannot exhibit oscillatory behaviour and their stability can be investigated using a natural Lyapunov function that can always be found by integration.

An arbitrary dynamic system can be represented as the sum of a gradient system and a conservative system. An analysis of the system in its decomposed form is shown to lead to a Lyapunov-like algorithm. The algorithm can be applied to yield directly a graphical indication of the qualitative behaviour of nonlinear second-order systems.

11.1 Gradient systems

Let $\phi: \mathsf{R}^n \to \mathsf{R}^1$ be a continuously differentiable function. Each $x \in \mathsf{R}^n$ for which $\partial\phi/\partial x_i = 0, \forall i \leq n$ is called a *critical point* of ϕ. Every other point is called a *regular point*.

A system of equations that can be written in the form

$$\dot{x}_i = f_i(x) = -\frac{\partial\phi}{\partial x_i}, \quad i = 1, \dots, n,$$

is called a *gradient system*. (The minus sign is traditional.) A system is a gradient system if and only if

$$\frac{\partial f_i}{\partial x_j} = \frac{\partial f_j}{\partial x_i}, \quad \forall i, j \leq n.$$

For $n = 3$ this condition is that curl $(f) = 0$ and the vector field for f is *irrotational*.)

A gradient system is *non-degenerate* if all its critical points are *Morse* (i.e., if at each critical point the Hessian matrix of ϕ is non-singular).

Let Σ be a non-degenerate gradient system and let ϕ attain its minimum value at a critical point $\bar{x} \in \mathbb{R}^n$. Define $V(x) = \phi(x) - \phi(\bar{x})$ then:

(a) $V(\bar{x}) = 0$;
(b) There exists a neighbourhood U of x in which $V(x) > 0$ and where $V(x)$ is strictly increasing. (This is so because $\bar{x}$ is a non-degenerate critical point and, for instance, a local coordinate system can be defined in U with $\bar{x}$ as the origin.)
(c) $dV/dt < 0$ everywhere in $\mathbb{R}^n$, except at critical points.

Proof:

$$\frac{dV}{dt} = \frac{d\phi}{dt} = \langle \nabla \phi, \dot{x} \rangle = \langle \nabla \phi, - \nabla \phi \rangle$$

$$= - |\nabla \phi|^2 < 0.$$

From (a), (b), (c), by Lyapunov's direct method, solutions are asymptotically stable from the neighbourhood U to the critical point $\bar{x}$. If the system is linearised near to $\bar{x}$, the Jacobian matrix of ϕ becomes the system matrix for the linearised system. Since $J = [\partial f_i / \partial x_j]$ is symmetric, the linearised system has only real eigenvalues and mutually orthogonal eigenvectors.

The solutions of a gradient system travel in the direction of the vector $\dot{x} = - \nabla \phi$, i.e., at all times orthogonally to the family of surfaces satisfying $\phi = $ constant. This property allows trajectories to be sketched in the phase plane.

11.2 Example of a gradient system

A simple example of a gradient system is given by

$$\left.\begin{array}{l} \dot{x}_1 = - 2x_1 (1 + x_2^2), \\ \dot{x}_2 = - 2x_2 (1 + x_1^2). \end{array}\right\} \tag{11.1}$$

The function ϕ is found from eqn. (11.1) by integration to be

$$\phi = x_1^2 + x_2^2 + x_1^2 x_2^2,$$

$$\dot{\phi} = 2x_1 \dot{x}_1 + 2x_2 \dot{x}_2 + 2x_1 x_2^2 \dot{x}_1 + 2x_1^2 x_2 \dot{x}_2$$

$$= 2x_1(-2x_1(1 + x_2^2)) + 2x_2(- 2x_2(1 + x_1^2))$$

$$+ 2x_1 x_2^2(- 2x_1(1 + x_2^2)) + 2x_1^2 x_2(- 2x_2(1 + x_1^2)) < 0$$

provided that $x_1, x_2 \neq 0$.

Thus ϕ is a Lyapunov function and the system is globally asymptotically stable to the critical point at $(0, 0)$.

11.3 Decomposition of second-order systems — introduction

(This approach is due to Evans and is described fully in Abd-Ali and Evans (1975a, 1975b).)

From experience of electrical network equations, we know that stability depends on the presence of dissipative components while periodic behaviour depends on the presence of energy storage components.

Let Σ be a linear time-invariant system with system matrix A. A can be decomposed into a symmetric matrix A_+, governing system stability and an asymmetric matrix A_-, governing system periodic behaviour.

The behaviour of the system Σ can be considered to take place in phase space in response to a force field. The symmetric matrix A_+ represents a gradient system with irrotational vector field that can be considered derived from a scalar field ϕ that always exists. Further, for a gradient system, ϕ is a natural Lyapunov function. Now, since all the stability information concerning Σ is in A_+ and not in A_-, it follows that ϕ might be a natural Lyapunov function for the system Σ, although Σ is not a gradient system. The loop of reasoning can be closed via Helmholtz theorem, which says that under certain not very restrictive conditions, any vector field can be decomposed into an irrotational field and a solenoidal field.

Recall from Chapter 2 that a system Σ is called a conservative system if its equation

$$\begin{pmatrix} \dot{x}_1 \\ \vdots \\ \dot{x}_n \end{pmatrix} = \begin{pmatrix} g_1(x_1, \ldots, x_n) \\ \vdots \\ g_n(x_1, \ldots, x_n) \end{pmatrix}, \quad \text{or in vector form} \quad \dot{x} = g(x),$$

satisfies the condition

$$\sum_{i=1}^{n} \frac{\partial g_i(x)}{\partial x_i} = 0.$$

For a three-dimensional system the condition is that div $g(x) = 0$. In fluid dynamic terms, the flow of such a system is incompressible or volume preserving and contains no sources nor sinks. The vector field for Σ is said to be *solenoidal* and there exists another vector field E such that

$$g(x) = \text{curl } E.$$

If a general vector field f is decomposed into a gradient system h and a conservative system g so that

$$f = h + g$$

then

$$\text{div } f = \text{div } h + \text{div } g = \text{div } (-\nabla \phi) = \nabla^2 \phi$$

for some potential function ϕ. This is Poisson's equation.

(Note that our definition of a conservative system follows well established practice as, for instance, in Minorsky (1947). This should not be confused with the concept of a conservative vector field, defined as a field in which every closed line integral is zero.)

11.4 The decomposition method for second-order nonlinear systems

Let a system Σ be decomposed into a gradient system and a conservative system. The system equations can be considered to be in the general form

$$\dot{x} = - \nabla \phi + G(x).$$

The function $\phi: R^2 \to R^1$ may or may not be a Lyapunov function for the system Σ. To investigate we examine

$$\frac{d\phi}{dt} = \langle \dot{x}, - \nabla \phi \rangle.$$

(Recall the geometric concept that the projection of $\dot{x}$ onto $\nabla \phi$ is given by

$$\left\langle \dot{x}, \frac{\nabla \phi}{\|\nabla \phi\|} \right\rangle \frac{\nabla \phi}{\|\nabla \phi\|}$$

which implies that trajectories move in the direction of decreasing ϕ if $d\phi/dt$ is negative.)

Define $\sigma = \langle \dot{x}, \nabla \phi \rangle$. Now clearly if both ϕ and σ are positive definite everywhere in the phase plane then by Lyapunov, the system Σ is stable.

The following two theorems allow the local behaviour of a decomposed system to be determined.

11.4.1 Theorem 11.1

Let $\bar{x}$ be a critical point at which ϕ takes on a local minimum value. If $\bar{x}$ lies in a region where $\sigma > 0$, $\bar{x}$ is a stable critical point.

Proof: Let U be a region around $\bar{x}$ in which $\sigma > 0$.

Let $x \in U$ be any point and let $\pi^+(x, t)$, $t > t_0$ be the half trajectory through x.

$$\lim_{\epsilon \to 0} \pi^+ \frac{(x, t_0 + \epsilon) - \pi^+(x, t_0)}{\epsilon} = \dot{x}(t_0).$$

The projection of $\dot{x}(t_0)$ onto $\nabla \phi$ is given by

$$g = \left\langle \dot{x}(t_0), \frac{\nabla 0}{\|\nabla \phi\|} \right\rangle \frac{\nabla \phi}{\|\nabla \phi\|},$$

since $\sigma = - \langle \dot{x}, \nabla \phi \rangle$ and $\sigma > 0$ then g is in the direction of decreasing ϕ.

11.4.2 Theorem 11.2

Let x^0 be a regular point on the trajectory $\pi^+(x^0, t)$ and let $\sigma > 0$ in a neighbourhood U of x^0, then a point on the trajectory $\pi^+(x^0, t)$ moves in the direction of decreasing ϕ in the neighbourhood U.

Proof: Exactly similar to that for the previous theorem.

11.5 Examples

Sometimes the required decomposition for the computation of ϕ is elementary. On other occasions an iterative technique (Leigh and Ng, 1980) needs to be used. Headland (1982) has obtained a closed expression to replace the iterative technique

A PDP11 computer was set up to perform the decomposition and plot ϕ, σ contours directly on a visual display.

The first example given is of a linear system to allow appreciation of the method. The second example gives an indication of the power of the approach for the graphical analysis of a nonlinear system.

11.5.1 Example 1 — Linear system with a stable node

$$\dot{x} = \begin{pmatrix} -2 & -1 \\ -2 & -3 \end{pmatrix} x = \begin{pmatrix} -2 & -1\cdot5 \\ -1\cdot5 & -3 \end{pmatrix} x + \begin{pmatrix} 0 & 0\cdot5 \\ -0\cdot5 & 0 \end{pmatrix} x,$$

$$\phi = x_1^2 + 1\cdot5x_1x_2 + 1\cdot5x_2^2,$$

$$\sigma = 7x_1^2 + 15\cdot5x_1x_2 + 10\cdot5x_2^2,$$

$\sigma > 0$ everywhere and the ϕ contours are very simple (Fig. 11.1). As expected, trajectories approach the origin to give an asymptotically stable system.

11.5.2 Example 2

$$\dot{x} = \begin{pmatrix} -1 & 2x_1^2 \\ 0 & -1 \end{pmatrix} x.$$

This yields on decomposition

$$\dot{x} = \begin{pmatrix} -1 & 2x_1^2 \\ 2x_1^2/3 & -1 \end{pmatrix} x + \begin{pmatrix} 0 & 0 \\ 2x_1^2/3 & 0 \end{pmatrix} x,$$

$$\phi = x_1^2/2 - 2x_1^3x_2/3 + x_2^2/2$$

$$\sigma = x_1^2 - 14x_1^3x_2/3 + 4x_1^4x_2 + x_2^2.$$

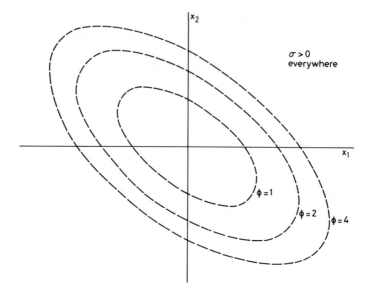

Fig. 11.1   φ *contours for the system of Example 1*

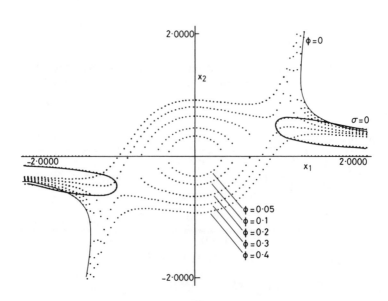

Fig. 11.2   φ, σ *contours for the system of Example 2*

This information is plotted in Fig. 11.2. It is clear that there is a stability region inside the $\phi = 0.2$ contour and that large regions in the second and fourth quadrants are also stable to the origin. These conclusions are confirmed by Lyapunov analysis. Taking

$$V = x_2^2 + x_1^2/(1 - x_1 x_2),$$

Which is positive definite for $x_1 x_2 > 1$ and

$$\frac{dV}{dt} = - 2x_1^2 - 2x_2^2,$$

which is negative definite.

Chapter 12

References

ABD-ALI, A., FRADELLOS, G. and EVANS, F. J. (1975a): 'Structural aspects of stability in nonlinear systems, Part I', *Int. J. Control*, **22**, No. 4, pp. 481–491.

ABD-ALI, A. and EVANS, F. J. (1975b): 'Structural aspects of stability in nonlinear systems', Part II, *Int. J. Control*, **22**, No. 4, pp. 493–516.

AGGARWAL, J. K. (1972): *Notes on Nonlinear Systems*, Van Nostrand, New York.

APOSTOL, T. M. (1974): *Mathematical analysis*, Addison-Wesley, New York.

BASS, R. W. (1961): 'Mathematical legitimacy of equivalent linearisation by describing functions', *Automatic and Remote Control, Proc. First International Congress of the International Federation of Automatic Control*, Vol. II, Butterworths, London, pp. 895–905.

BOOTON, R. C. (1954): 'Nonlinear control systems with random inputs', *Trans. IRE, PGCT*, **CT-1**, pp. 9–18.

BRAUN, M. (1975): *Differential Equations and Their Applications*, Springer, New York.

DESOER, C. (1965): 'A generalisation of the Popov criterion', *IEEE Trans.* **AC-10**, No. 2, pp. 182–185.

DRIVER, R. D. (1977): *Ordinary and Delay Differential Equations*, Springer, New York.

GRENSTED, P. E. W. (1962): 'Frequency response methods applied to nonlinear systems', in *Progress in Control Engineering*, Volume 1, Heywood, London.

HEADLAND, L. (1982): 'Comments on an algorithm for decomposing a two dimensional vector field,' *IEEE Trans. Auto. Control* **AC-27**, No. 2.

HIRSCH, M. W. and SMALE, S. (1974): *Differential Equations, Dynamical Systems and Linear Algebra*, Academic Press, New York.

KALMAN, R. E. and BERTRAM, J. R. (1960): 'Control system analysis and design via the second method of Lyapunov', *J. Basic Engineering*, **82**, Jeriso D, No. 2, pp. 371–393.

KOCHENBURGER, R. J. (1950): 'A frequency response method for analysing and synthesising contactor servomechanisms', *Trans. IEEE*, **69**, No. 1, pp. 270–284.

KREYSZIG, E. (1974): *Advanced Engineering Mathematics*, 4th Edition, John Wiley, New York.

KRYLOV, N. M. and BOGOLIUBOFF, N. N. (1932): 'Quelques examples d'oscillations non linéaires', *C. R. Acad. Sci. Paris*, **194**, p. 57.

LA SALLE, J. P. and LEFSCHETZ, S. (1961): *Stability by Lyapunov's Direct Method with Applications*, Academic Press, New York.

LEFSCHETZ, S. (1977): *Differential Equations — Geometric Theory*, Dover, New York.

LEIGH, J. R. and LI, C. K. (1981): 'An adaptive control method with application to ship steering', *IEE International Conference on Control and its Applications*, Coventry, England, March 1981.

LEIGH, J. R. and NG, M. H. (1980): 'An algorithm for decomposing a two-dimensional nonlinear vector field', *IEEE Trans. Auto Control* **AC-25**, No. 1, pp. 131–132.

LEIGH, J. R. and NG, M. H. (1981): 'Correction to "An algorithm for decomposing a two-dimensional nonlinear vector field,"' *IEEE Trans. Auto Control* **AC-26**, No. 2, p. 618.

LYAPUNOV, A. M. (1966): *Stability of Motion*, Academic Press, New York.

MINORSKY, N. (1947): *Nonlinear Mechanics*, J. S. Edwards, U.S.A.)

POINCARÉ, H. (1928): *On Curves Defined by a Differential Equation*, Gauthier-Villars, Paris.

POPOV, V. M. (1961): 'Absolute stability of nonlinear sytems of automatic control', *Automation and Remote Control*, **22**, No. 8, pp. 857–875.

SANCHEZ, D. A. (1968): *Ordinary Differential Equations and Stability Theory: An Introduction*, W. H. Freeman and Co., San Francisco.

SHINNERS, S. M. (1972): *Modern Control System Theory and Application*, Addison-Wesley, Reading, Mass.

TSYPKIN, JA. Z. (1958): 'Relations between the coefficient of equivalent gain and the gain characteristic', *Regelungstechnik* **8**, p. 285 (In German).

TUSTIN, A. (1947): 'The effects of backlash and speed dependent friction on closed cycle control systems', *J. IEE* **94**, Part 2A, p. 143.

WILLEMS, J. L. (1970): *Stability Theory of Dynamical Systems*, Nelson, London.

ZADEH, L. A. (1956): 'On the identification problem', *Institute of Radio Engineers Group on Circuit Theory*, **CT-3**, p. 277.

ZAMES, G. (1966): 'On the input–output stability of time-varying nonlinear feedback systems, Parts I and II, *IEEE Trans.* **AC-11**, No. 2, pp. 228–238 and No. 3, pp. 465–476.

Chapter 13

Bibliography

13.1 Further historical and supporting literature

ANDRANOV, A. A., VITT, A. A. and CHAIKIN, S. C. (1966): *Theory of Oscillations*, Pergamon Press, Oxford.

ANDRANOV, A. A. (1973): *Qualitative Theory of Second Order Dynamic Systems*, John Wiley, New York.

BENDIXSON, L. (1901): 'Sur les Courbes Définies par des Équations Différentielles', *Acta Mathematica*, **24**, pp. 1–88.

BOGOLIUBOV, N. N. and MITROPOLSKY, Y. A. (1961): *Asymptotic Methods in the Theory of Nonlinear Oscillations*, Hindustan Publishing Corporation, India.

CESARI, L. (1963): *Asymptotic Behaviour and Stability Problems in Ordinary Differential Equations*, Academic Press, New York.

HADDAD, A. H. (ed.) (1975): *Nonlinear Systems*, Halsted Press, U.S.A.

HAYASHI, C. (1964): *Nonlinear Oscillations in Physical Systems*, McGraw-Hill, New York.

KRYLOV, N. M. and BOGOLIUBOV, N. N. (1943): *Introduction to Nonlinear Mechanics*, Princeton University Press.

LEFSCHETZ, S., 1(1950), 2(1952), 3(1956), 4(1958), 5(1960): *Contributions to the Theory of Nonlinear Oscillations*, Princeton University Press.

VAN DER POL, B. (1934): 'Nonlinear theory of electric oscillations', *Proc. IRE* **22**, pp. 1051–1086.

STOKER, J. J. (1950): *Nonlinear Vibrations*, Interscience Publishers, New York.

13.2 Theoretical foundations

The literature in this section indicates the possibility of applying topological, function-analytic and differential geometric concepts and methods to the study of nonlinear systems. It appears, in particular, that aspects of differential topology and degree theory as presented by Guillemin (1974) and Lloyd (1978) could be used to link with and extend established concepts of nonlinear control theory.

Readers interested to pursue these aspects could perhaps start by familiarising themselves with the following concepts and their inter-relations: The *degree* of a mapping, *intersection number, Euler characteristic, Lefschetz number* and the *Poincaré–Hopf theorem*.

AGAUBAEV, K. SH., BARKIN, A. I. and POPKOV, YU. S. (1976): 'Analysis of nonlinear systems described by Volterra series', *Autom. and Remote Control (U.S.A.)*, **37**, No. 11, pp. 1631–1640.

BARRETT, J. F. (1963): 'The use of functionals in the analysis of nonlinear physical systems', *J. Electronics and Control* **15**, No. 6, pp. 567–615.

BROCKETT, R. W. (1976): 'Nonlinear systems and differential geometry', *Proc. IEEE*, **64**, pp. 61–72.

GUILLEMIN, V. and POLLACK, A. (1974): *Differential Topology*, Prentice-Hall, New York.

ISIDORI, A., KRENER, A. J., GORI-GIORGI, C. and MONACO, S. (1981): 'Nonlinear decoupling via feedback: a differential geometric approach', *IEEE Trans. Autom. Control (U.S.A.)*, **AC-26**, No. 2, pp. 331–345.

LLOYD, N. G. (1978): *Degree Theory*, Cambridge University Press.

SAEKS, R. (1978): 'An index theory for nonlinear systems', *Proc. 1978 IEEE International Symposium on Circuits and Systems, New York*, 17–19 May 1978, p. 503.

SELL, G. R. (1971): *Topological Dynamics and Ordinary Differential Equations*, Van Nostrand, New York.

13.3 Multivariable and discrete time systems

CHEN, T. C., HAN, K. W. and THALER, G. J. (1981): 'Stability analysis of multirate nonlinear sampled data control systems', *IEE Int. Conf. on Control and its Applications*, Coventry, England, March 1981.

COOK, P. A. (1979): 'Nonlinear multivariable systems', in *Modern Approaches to Control System Design* (Edited by N. Munro), Peter Peregrinus.

DILIGENSKI, S. N. (1979): 'Estimation of the attraction region of multi-dimensional nonlinear systems by aggregation–decomposition method', *Autom. and Remote Control (U.S.A.)*, **40**, No. 5, pp. 24–37 and pp. 645–658.

GORDEEV, A. A. (1977): 'Approach for analysing nonlinear systems of high order on many-sheet phase plane', *Autom. and Remote Control (U.S.A.)*, **38**, No. 12, pp. 1872–1875.

JURY, E. I. and LEE, B. W. (1964): 'On the absolute stability of nonlinear sampled-data systems', *IEEE Trans*. **AC-9**, No. 4, pp. 551–554.

KACZOREK, T. (1981): 'Deadbeat control in multivariable nonlinear time-varying systems', *Int. J. Syst. Sci. (GB)*, **12**, No. 4, pp. 393–405.

MILLER, B. M. (1978): 'Nonlinear sampled data control of processes described by ordinary differential equations, I and II, *Autom. and Remote Control (U.S.A.)*, **39**, No. 1, Part 1, pp. 57–67 and No. 3, Part 1, pp. 338–344.

PECZKOWSKI, J. L. and STOPHER, S. A. (1980): 'Nonlinear multivariable synthesis with transfer functions', *1980 Joint Automatic Control Conference, San Francisco, U.S.A.*, 13–15 August 1980.

REZTSOV, V. P. (1977): 'A method of investigation of nonlinear sampled data systems by Lyapunov's direct method', *Sov. Autom. Control (U.S.A.)*, **10**, No. 4, pp. 63–65.

SHARMA, T. N. and VIMAL SINGH (1981): 'On the absolute stability of multivariable discrete time nonlinear systems', *IEEE Trans. Autom. Control (U.S.A.)*, **AC-26**, No. 2, pp. 585–586.

SINGH, S. N. (1980): 'Decoupling of invertible nonlinear systems with state feedback and precompensation', *IEEE Trans. Autom. Control (U.S.A.)*, **AC-25**, No. 6, pp. 1237–1239.

SOMMER, R. (1980): 'Control design for multivariable nonlinear time-varying systems', *Int. J. Control (GB)*, **31**, No. 5. pp. 883–891.

VIDAL, P. (1972): *Nonlinear Sampled Data Systems*, Gordon and Breach, New York.

13.4 Other control theory

CHONG-HO CHOI (1980): 'Identification of nonlinear discrete systems in the time domain', *J. Korean Inst. Electr. Eng.* **29**, No. 11, pp. 742–750.

FAKUMA, A. (1978): 'Jump resonance in nonlinear feedback systems — approximate analysis by the describing function method', *IEEE Trans. Autom. Control (U.S.A.)*, **AC-25**, No. 5, pp. 891–896.

HIRAI, K. and SAWAI, N. (1978): 'A general criterion for jump resonance of nonlinear control systems', *IEEE Trans. Autom. Control (U.S.A.)*, **AC-25**, No. 5, pp. 896–900.

HIRAI, K., IWAI, M. and USHIO, T. (1981): 'Catastrophic jump phenomena in a nonlinear control system', *IEEE Trans. Autom. Control (U.S.A.)*, **AC-24**, No. 2, pp. 601–603.

JACOBS, O. L. R. (1980): 'On nonlinear adaptive control', Report Number 1323/80, Oxford University Engineering Laboratory (GB).

PETROV, B. N., VASILYEV, V. I. and GUSEV, YU. M. (1980): 'Synthesis of nonlinear automatic control systems by the method of generalised linearisation', *Eng. Cybern. (U.S.A.)*, **18**, No. 1, pp. 111–116.

POLAK, E., MAYNE, D. Q. (1980): 'Design of nonlinear feedback controllers', *Proc. 19th IEEE Conf. on Decision and Control, Albuquerque, New Mexico*, Volume 2, pp. 1100–1104.

PORTER, B. (1980): 'Necessary conditions for absolute stability of nonlinear regulators', *Electron. Lett. (GB)*, **16**, No. 5, pp. 181–182.

POSTNIKOV, N. S. and SABAEV, E. F. (1980): 'Matrix comparison systems and their applications to automatic control problems', *Autom. and Remote Control (U.S.A.)*, **41**, No. 4, Part 1, pp. 455–464.

ROHELLA, R. S. and CHATTERJEE, B. (1980): 'Effect of time delay on nonlinear systems', *J. Inst. Electron and Telecommun. Eng. (India)*, **25**, No. 9, pp. 386–388.

SIDOROV, I. M. and TIMOFEYEV, V. V. (1979): 'Analysis of nonlinear automatic control systems in a resonance region', *Eng. Cybern. (U.S.A.)*, **17**, No. 3, pp. 128–135.

Special issue on nonlinear systems (1981): *Proc. IEE*, **128**, Part D, No. 5, Sept. 1981.

13.5 Selected application papers

BASHARIN, A. V., BASHARIN, I. A. and BELOV, A. V. (1979): 'A computer method for determining self-excited oscillations and their parameters in nonlinear control systems (servo drives), *Sov. Electr. Eng. (U.S.A.)*, **50**, No. 9, pp. 50–56.

BROADWATER, R. P. (1980): 'Application of nonlinear disturbance-accommodating controllers in power plant control system design', *Proc. 19th IEEE Conf. on Decision and Control, Albuquerque, New Mexico*, Volume 2, pp. 107–112.

HOROWITZ, I., GOLUBOV, B. and KOPELMAN, T. (1980): 'Flight control design based on nonlinear model with uncertain parameters', *J. Guid. and Control (U.S.A.)*, **3**, No. 2, pp. 113–118.

LAKOTA, N. A., LISENKO, S. A., RAKMANOV, YE. V. and SHVEDOV, V. N. (1979): 'Design of servomechanisms of manipulating robot for precise trajectory realisation', *Eng. Cybern. (U.S.A.)*, **17**, No. 5, pp. 32–37.

LETHERMAN, K. M. (1980): 'Analytical methods for thermostatic on-off controls', *Build. Serv. Eng. Res. and Technol. (GB)*, **1**, No. 1, pp. 31–34.

Index